p. 98 problems with pickles-

# *MODERN FOOD PRESERVATION*

# *MODERN FOOD PRESERVATION*

*By*

*Margaret McWilliams*
*California State University, Los Angeles*

*and*

*Harriett Paine*
*Cerritos College*

*PLYCON PRESS*
*1916 North Gilbert, Fullerton, CA 92633*

*Library of Congress Catalog Card Number 77-76339*

**ISBN 0-916434-25-7**

10 9 8 7 6 5 4 3 2

Printed in the United States of America

# Preface

Food preservation, in its various forms, has undergone a resurgence recently as a result of several influences on lifestyles today. The interest in the heritage of our past has prompted many people to undertake kitchen projects which had almost been relegated to history books, and the pleasure of baking bread, gardening, and preserving foods is being rediscovered. Concern over the safety of our food supply has motivated some consumers to preserve and prepare their own foods rather than relying so heavily on industry for prepared and convenience food items. The spiraling costs of food in today's markets provide additional impetus to the desire to preserve food when supplies are high and costs are correspondingly low. The need for a reserve food supply to meet emergency situations stimulates others to preserve food at home. These are but some of the reasons why consumers are drying, canning, pickling, freezing, and preserving foods at home in ever-increasing numbers.

Home preserved foods can provide satisfactions which go beyond the immediate nutritional contribution they make. There can be significant savings in food costs when food preservation is managed carefully. A reserve supply of food prepared at home provides the psychological satisfaction of being self-sufficient. There also can be the pleasure of making specialty items which are enjoyed by family and friends, but which are not available in the same quality from commercial sources. The choice of the method of food preservation to be used is purely an individual matter; practically all foods can be preserved by drying, canning, or freezing. However, the characteristics of the finished product will vary with the method of preservation selected. Preferences will be identified as the different methods of food preservation are explored.

Of particular importance in any method of preserving foods is the need for insuring that the preserved food is safe to eat and that it is of optimum palatability. This book provides valuable insights into the microbiological aspects of food safety for each of the methods of preservation. The principles underlying the preservation methods are reviewed in detail to provide a sound foundation for each type of preservation. The structural features of plant foods are outlined to provide greater understanding of the changes which occur when foods are dried, canned, or frozen. By providing the appropriate scientific basis of preservation techniques, this book will aid consumers in preserving foods safely and palatably. The unique concerns associated with each type of preservation are discussed in detail. The extensive illustrations have been planned to aid in achieving the goal of preserving foods of excellent quality at home.

It is our hope that this book will foster a growing interest in the various methods of food preservation which can be done at home. In addition, we have presented the scientific foundations of food preservation so that the products preserved in the home will be handled to assure safety and excellent quality. Use of proper food preservation techniques is one way of helping to assure that optimum use will be made of available food supplies. This is an important step which individuals and families can take to contribute toward solving the world's food supply problems.

For their assistance and encouragement in the preparation of this book, we wish to acknowledge Dr. George York of the University of California at Davis, Marjorie Stevenson of the Cooperative Extension Service at the University of Nevada, Reno, and Mary Kramer of California State University, Los Angeles.

Margaret McWilliams
Harriett Paine

Los Angeles, California
July, 1977

# Table of Contents

# Photographic Credits

**ii Ball Corporation**
**viii Ball Corporation**
Chapter 1. California Raisin Advisory Board
Fig. 1.1. California Raisin Advisory Board
Fig. 1.2. General Foods Corp.
Chapter 3. California Plum Commodity Committee
Fig. 3.1. California Raisin Advisory Board
Fig. 3.33. U.S. Dept. of Agriculture
Chapter 4. California Plum Commodity Committee
Fig. 5.23. Mirro Aluminum Co.
Chapter 10. Wheat Flour Institute
Chapter 11. J. Walter Thompson
Fig. 11.1. J. Walter Thompson
Fig. 11.2. United Fresh Fruit and Vegetable Assoc.
Fig. 11.8. Burpee Seeds
Fig. 11.14. U.S. Dept. of Agriculture
Fig. 11.15. U.S. Dept. of Agriculture
Fig. 11.16. U.S. Dept. of Agriculture
Fig. 11.17. U.S. Dept. of Agriculture
Fig. 11.18. National Turkey Federation
Fig. 11.19. Green Giant Co.
Fig. 11.20. Betty Crocker Kitchens of General Mills
Fig. 11.21. J. Walter Thompson
Fig. 11.22. American Egg Board

# Chapter 1. Historical Perspective

## INTRODUCTION

For many centuries, food preservation has been an important aspect of human existence, but it has only been in the past 175 years that many variations in the processes have developed. Prior to the pioneer heat-processing work of Nicolas Appert in France around the beginning of the nineteenth century, drying had been the technique utilized for preserving. The concept of preserving by the use of heat opened new avenues, not only for home preservation, but also for subsequent commercial food processing. More than a hundred years later the availability of frozen storage, both in the home and commercially, made it possible to add freezing to the techniques which could be used for preserving food. Present-day technology has added irradiation and freeze-drying to the methods suited to commercial preservation, but the home methods remain the basic methods of drying, canning (including preserving and pickling), and freezing.

### Drying

The origins of food preservation by drying are lost in antiquity, for seemingly this technique precedes recorded history. References to dried foods even are included in the Old Testament of the Bible:

> He shall separate himself from wine, and strong drink, and shall drink no vinegar of wine, or vinegar of strong drink, neither shall he drink any liquor of grapes, nor eat moist grapes, or dried.
>
> *Numbers 6:3*

> Then Abigail made haste, and took two hundred loaves, and two bottles of wine, and five sheep ready dressed, and five measures of parched corn, and an hundred clusters of raisins, and two hundred cakes of figs, and laid them on asses.
>
> *I Samuel 25:18*

Clearly, this method of preserving food was practiced commonly in the Mediterranean regions at that time.

Sun drying of foods, such as the corn, grapes, and figs cited in the Bible, was a method known and practiced by the ancient Egyptians, Persians, and Greeks. It is thought that, perhaps as long as 3000 years ago, potatoes were dried to make "chuño", an important food in the diets of natives living in the Andes of South America at that time. Historians think that chuño may represent the first food stored by man to provide the food needed to survive the difficult period of winter and spring.

Pre-Columbian American Indians are credited with the development of pemmican. This food was prepared by air drying buffalo meat or venison mixed with fat. The resulting product could be stored or carried easily without spoiling for many months, thus permitting regular use of meat in the diet rather than necessitating a "feast or famine" approach to hunting. This food item became an important part of the food supply of trappers in the early days of exploration and settling of the North American continent.

Drying of meat was known in Europe for centuries. Homer mentioned the use of dried meats on voyages. The drying of meats in Europe involved not only preservation by dehydration, but also the added role of salts as an aid in curing. Salted, dried meats were prominent in Viking explorations, as

*Figure 1.1. Sun drying, although an ancient method of food preservation, is still practiced today commercially as well as in the home. It has the advantage of easy storage, as well as being light in weight.*

well as in the voyages of other Europeans. Of course, these dried products had the advantage of being relatively light in weight as well as being safe to eat.

Yet another example of dried foods in history can be found in accounts of Marco Polo's travels in the 13th century. He noted the use of sun-dried milk by the Mongols he encountered. By 1810, dried milk had been prepared in a tablet form by Nicolas Appert.

In the 18th century, artificial drying of foods began to appear, although sun drying continued as a popular method of food preservation. The need for a reliable source of food for feeding army troops motivated this development. Of particular concern was the need for a food source to prevent scurvy among troops in the field. Of course, ascorbic acid had not been identified at that time, but there was recognition that some foods seemed to be useful in preventing scurvy. The early attempts to use dried vegetables in the field in feeding the British troops were not viewed as successful because of the problem with scurvy. The vegetables used to feed troops in the Crimea in the mid-19th century were dried artificially in a room heated by a stove, a process which destroyed the ascorbic acid in the vegetables. Subsequently, the value of scalding vegetables to retain their anti-scorbutic activity was discovered, and dried vegetables became nutritionally more important for troops. There was extensive use of dried vegetables by the British in the Boer War at the beginning of the present century, and dried products also were of importance in the First World War.

Largely due to the stimulus of wars, research was continued into the drying of foods. Numerous commercially feasible drying techniques evolved. Vacuum drying was developed as a technique for preserving many foods, including fish and meat. Tunnel drying emerged as a device for drying fruits; this device circumvented the problems created when untimely rains would interfere with sun drying of fruits. Drum drying proved to be a method particularly well suited to the drying of milk and milk products, as well as to tomatoes. Milk and eggs also were spray dried satisfactorily on a commercial basis. Skim milk and egg yolks were especially well suited to the spray drying process. Potatoes were another food which proved to be particularly well suited to commercial drying.

Although much of the commercial drying industry developed due to the stimulus of wartime needs, the industry has not withered during peacetime. The evolution of packaged mixes, ranging from dehydrated soups through puddings and other desserts, has maintained the demand and interest in dried foods commercially. Particularly outstanding markets have been found among consumers for nonfat dry milk solids and various dried potato products. The ease of storage and use have made these items become staples in the kitchens of many consumers today.

Home drying of foods had virtually disappeared until quite recently. However, the emphasis on ecology, energy conservation, and natural foods has triggered considerable interest in this form of food preservation. One of the motivating forces also is the need for safe, lightweight food supplies for camping and backpacking. Yet another factor is the concern with providing homes with emergency supplies as a buffer against blizzards, earthquakes, and other natural disasters which often leave hunger in their wake. Dried foods, whether prepared at home or commercially, have distinct advantages for people who are carrying, storing, and preparing their food far away from the supermarket and the refrigerator.

## CANNING

The use of heat in food preservation is a relatively new development when viewed on the continuum of man's recorded history, for it was not until the beginning of the 19th century that there is evidence that thermal preservation was done. The credit for this innovative approach toward making food available throughout the year is given to Nicolas Appert, a Frenchman who developed this technique in response to a French military plea for a safe way to preserve foods for use on land and sea by troops. For this pioneer research effort, Appert was awarded a prize of 12,000 francs in 1809; he also earned for himself the unofficial title of "Father of Canning". Appert's work is even more impressive when it is recognized that he had to develop the jars for processing, even to the point of fashioning and gluing the corks which were used as closures. Thus from the needs of the military was canning as a means of food preservation begun. Since Appert's time, the method has been a key process for handling large food crops, both in the home and in industry. The canned food industry remains a very large segment of the commercial food world, despite challenges from the newer methods of freezing and freeze-drying.

A significant factor in the development of canning as a means of food preservation was the work done by Louis Pasteur and others in the field of food microbiology. It was through the experiments of such researchers as Samuel Prescott, W. Lyman Underwood, and Pasteur that science entered the vital realm of food preservation. Their work was followed by developments in the designs and materials used in the containers themselves. Then in 1858, John L. Mason invented the canning jar which still bears his name. The Mason jar involved development of a bottle with a threaded neck which would fit with the threads in a metal cap to screw down tightly on a rubber gasket, thus effecting an excellent seal. This type of closure still is in use today, although it has been augmented with various other closure designs. Scientists began work utilizing the thermocouple to measure the heat curves in processing and to determine much more accurately the heat needed for safe processing of various foods.

In the present century, as the research on vitamins gained momentum, the effect of heat on vitamins began to be of concern. The observation that higher temperatures for shorter periods of time were less destructive to vitamins than lower temperatures for longer processing times influenced the canning industry's approach to heat processing. The key to commercial canning became HTST (High-Temperature Short-Time) processing. Developments from this point in the commercial canning industry included agitation during cooking to facilitate uniformly high heat distribution throughout the food and aseptic methods of canning. Interest also began to be directed toward the problems identified with heat-resistant enzymes, for enzymes as well as microorganisms could cause deleterious changes during storage of canned foods.

As the food industry mobilized its efforts and accelerated its progress in the use of heat to preserve foods, the home canning of foods began to decline. Homemakers in many instances were beginning to be farther removed from inexpensive, abundant sources of food as the populations moved from the rural to the urban life in ever-increasing numbers. The practicality of home canning for many people began to be a question. In addition, the quality of commercial products was improving, and the cost was competitive. Thus, for many people home canning was relegated to a booth at the county fair, but was not a part of their daily lives.

However, recently this trend has been reversed. Interest in home food preservation in all forms has been increasing once again. In fact, in the summer of 1975 there was such a resurgence that the container industry was unable to provide enough of the closures for canning jars to meet the demand, and interest appears to be remaining high. This renewed focus on home food preservation seems to be occurring in tandem with renewed efforts at home gardening; these two obviously are very compatible endeavors.

## FREEZING

Although the effectiveness of frozen storage for food has been known for centuries to those persons living in areas where long, severe winters made such storage possible, the widespread use of freezing as a preservation technique had to wait until the commercial and home refrigeration units had been developed to the point of being reliable and economically feasible. As a result, freezing of food is largely a product of the present century, although the first patent noted in this field was awarded in 1842 to H. Benjamin in England. His patent was for a process of freezing by immersing in an ice-salt brine solution. Interest at that time was directed primarily toward the freezing of fish and meats, which are obviously very perishable items when stored under the conditions generally available in the 19th century. The first recorded use of freezing during transit as a means of keeping foods safe to eat apparently was an accident. A shipment of meat was sent from Australia to England in 1880; the ship was refrigerated, but the conditions were such that the meat became frozen and was shipped in that state. The results were viewed as being so favorable that this triggered the interest in the shipment of frozen, rather than refrigerated foods.

The development of frozen fruits apparently began in Colorado in the first decade of the 20th century in response to a large crop of strawberries and cherries, and a need for a method to hold the fruit for later sale. The early freezing techniques involved rather slow freezing of the food, which caused considerable disruption of the cellular structure. A quick freezing method, that is, one in which the food could be frozen in only a few hours versus several days, was developed in Germany in 1916. From this innovation, work to improve freezing methods was accelerated. The pioneer in developing frozen foods commercially for the retail market was Clarence Birdseye in the United States. His use of small containers for the retail trade launched this phase of the industry.

Freezing as a method for preserving food in the home was dependent on the availability of frozen storage space. For many homemakers, this problem first was solved by renting locker space in a frozen food locker plant. This enabled families to freeze sides of beef, large harvests from their gardens, or other foods, such as the rewards of hunting and fishing. The disadvantage was the inconvenience of driving to the locker plant and transporting the food back to the kitchen. With the availability now of highly reliable home freezers, this disadvantage to home freezing has virtually disappeared.

The advantages of freezing foods appeal to many homemakers. It is a simple matter to freeze small quantities of food. No longer does food preservation need to occupy the kitchen and one's time for a day or more at a time. The quality of many foods remains excellent when frozen for later use. No special equipment (other than a place for frozen storage) is required, for the containers appropriate for freezing often already are found in the kitchen.

## SUMMARY

Food preservation has been important to man through the ages. Sun drying probably was the first method used. This method later has been joined by artificial methods of dehydration, both in the home and commercially. Heat processing is the basis of canning in the home and commercially, although the equipment and technology for industry have been developed to a far greater degree than for the home canning arena. Heat treatment is used for the preparation of pickles, jams, and jellies, as well as for "canned" goods. Freezing is the most recent method of food preservation to gain widespread use. Its development at home and commercially was dependent upon the availability of adequate storage and shipping facilities.

## BIBLIOGRAPHY


Burr, H.K. and R.M. Reeve. Potatoes. In *Food Dehydration*. Vol. 2. Ed., W.B. Van Arsdel, et al. Avi. Westport, Conn. P. 83. 1973.

Coulter, S.T. and R. Jenness. Dry milk products. In *Food Dehydration*. Vol. 2. Ed., W.B. Van Arsdel, et al. Avi. Westport, Conn. P. 290. 1973.

Feinberg, B. Vegetables. In *Food Dehydration*. Vol. 2. Ed., W.B. Van Arsdel, et al. Avi. Westport, Conn. P. 1. 1973.

Goldblith, S.A., et al. *Introduction to Thermal Processing of Foods*. Avi. Westport, Conn. 1961.

Nury, F.S., et al. Fruits. In *Food Dehydration*. Vol. 2. Ed., W. B. Van Arsdel, et al. Avi. Westport, Conn. P. 158. 1973.

Stumbo, C.R. *Thermobacteriology in Food Processing*. Academic Press. New York. 2nd ed. 1973.

Van Arsdel, W.B., et al. *Food Dehydration*. Vol. 2. Avi. Westport, Conn. 1973.

Van Arsdel, W.B., et al. *Quality and Stability of Frozen Foods*. Wiley-Interscience. New York. 1969.


*Figure 1.2. Freezing is a method of food preservation which is very convenient, but has the disadvantage of requiring special storage equipment.*

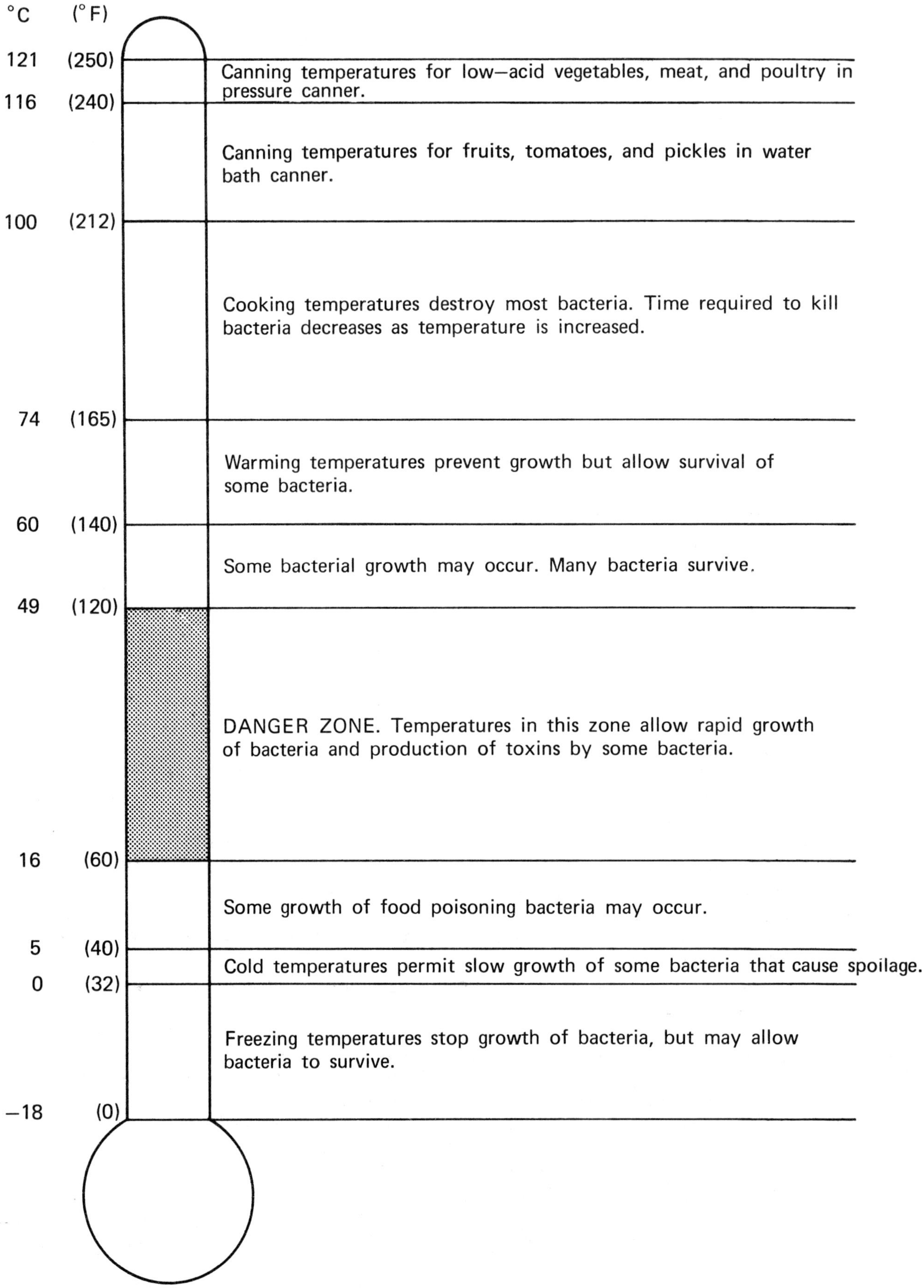
°C (°F)
121 (250)
Canning temperatures for low–acid vegetables, meat, and poultry in pressure canner.
116 (240)
Canning temperatures for fruits, tomatoes, and pickles in water bath canner.
100 (212)
Cooking temperatures destroy most bacteria. Time required to kill bacteria decreases as temperature is increased.
74 (165)
Warming temperatures prevent growth but allow survival of some bacteria.
60 (140)
Some bacterial growth may occur. Many bacteria survive.
49 (120)
DANGER ZONE. Temperatures in this zone allow rapid growth of bacteria and production of toxins by some bacteria.
16 (60)
Some growth of food poisoning bacteria may occur.
5 (40)
Cold temperatures permit slow growth of some bacteria that cause spoilage.
0 (32)
Freezing temperatures stop growth of bacteria, but may allow bacteria to survive.
–18 (0)

# Chapter 2. Microbiological Aspects

Microbiological contamination in preserved foods is of interest because of (1) effects on the storage time and the quality of the food following storage, and (2) possible health hazards associated with ingesting contaminated food. Methods of food preservation have been developed to minimize growth of microorganisms during the storage period, thus promoting longer shelf life and reduced hazard from eating the food. The problems associated with the various methods of food preservation vary with the method. These are considered in some detail in this chapter.

## POTENTIAL CONTAMINANTS

The microorganisms which may be found in preserved foods are of the same types as are found in fresh foods, notably molds, yeasts, bacteria, and toxins produced by these organisms. These organisms can enter the food supply via a number of routes, including the soil in which they are grown, the air around them, the water in which they may have been washed, or from food handlers. In short, the total environment prior to processing for preservation affords opportunities for contamination of foods.

Since there are so many opportunities for foods to become infected with microorganisms, attention should be directed toward minimizing these risks as much as possible. Reduction in the populations of microorganisms prior to the actual processing of foods to be preserved will be an aid in assuring the greater safety and shelf life of the processed product.

In commercial food processing, some microorganisms are cultivated for their desired effects on foods. Examples can be found in the production of cheeses and breads, as well as many other products. Controlled use of the microorganisms is the key to success in such foods. However, in home food preservation methods, the goal is to avoid microorganisms as much as possible. Therefore, the discussion in this chapter is directed primarily toward the elimination, or at least the reduction, of these populations.

The goals of food preservation, in relation to microbiological concerns, are to (1) minimize the original contamination of the food, (2) process the food in a manner designed to reduce the populations to an absolute minimum, and (3) store the processed food in a way suited to keep the growth of these populations during storage to as slow a rate as can be achieved.

Achievement of these goals requires careful attention to selecting, washing, and preparing the food for processing. The kitchen environment needs to be maintained with high levels of sanitation. Counters, cutting boards, and sinks must be kept meticulously clean. The food handler needs good habits of sanitation, including careful washing of hands and the avoidance of sneezing and handling of hair. Similarly, the food must be examined with great care so that any spoiled areas can be removed. Washing needs to be done very thoroughly to remove as much of the dirt, foreign materials, and insecticides as possible. These are the types of procedures which help to assure that the food entering the processing steps will carry with it an absolute minimum of contamination.

The processing steps need to be done carefully and accurately, with little time being wasted during the processing. Careful attention should be given to the timing of foods being blanched or heat processed. Accuracy in timing and reaching the correct temperatures are essential to destruction of potentially harmful microorganisms without overcooking the foods. Equipment, particularly that used for pressure canning, needs to be checked for accuracy so that the desired temperatures can be reached with precision.

Storage conditions also need to be considered. The integrity of packaging containers is of great importance in all types of storage, whether it be for canned, frozen, or dried foods. Breaks in the containers permit fresh contamination from external sources and will reduce the effective shelf life of the product. The tightness of the container also determines whether moisture can enter or leave the environment of the stored food.

## MICROORGANISMS AND DRIED FOODS

The general comments in the preceding section apply to the foods which are to be dried. Thus, yeasts, molds, and bacteria all may be present on the food. The drying process itself is generally destructive to yeast and to most bacteria. However, spores of bacteria and molds commonly survive the drying process. There also are a few heat-resistant bacteria which will not be destroyed by drying, even when some heat is applied in the dehydration process.

Means commonly employed to help reduce the potential population include the sulfuring of some fruits and vegetables and the blanching of vegetables prior to drying. Peeling of the produce can be another means of reducing contamination, but it only serves this function if clean equipment is used. Of course, removal of soft, spoiled areas is essential, too. Rinse water also needs to be meticulously clean if additional contamination is to be avoided.

With proper storage conditions after drying, microorganisms should not grow on the food. Bacteria generally do not grow in environments with less than 16 percent available moisture. Yeasts require a little higher moisture level (specifically, at least 20 percent) than most bacteria. Molds, with their ability to survive at moisture levels as low as 13 percent, are the most resistant to destruction during storage of dried foods. As may be expected, the microorganisms which are best able to survive the actual drying process also are the ones which will be most resistant to destruction during the storage period. However, there is continuing destruction of microorganisms during the storage of dried foods, with the greatest destruction occurring in the early part of the storage period and the rate of destruction slowly diminishing.

Repackaging of dried foods may lead to fresh contamination of the food; consequently, this practice is not encouraged. The practice of sweating some dry fruits also may cause increased growth of microorganisms because of the increased moisture content of the environment.

The temperature of the water used to rehydrate dried foods also influences the microorganismal count. For example, use of water at 50°C (122°F) to rehydrate chicken or turkey helps to reduce the count, compared with the use of cooler water, but water at simmering temperatures (between 85 and 100°C, that is 183 to 212°F) virtually eliminates growth. Thus, when feasible, the use of simmering water is recommended for rehydrating dried foods. This is particularly appropriate if there is a possibility of contamination with *Staphylococcus aureus*, since this bacterium can survive drying.

The bacteria on vegetables are likely to be greater in variety and number than those found on dried fruits, but the actual populations are quite dependent upon the handling of the food prior to and during drying. The bacteria which may be found in dried fruits and vegetables include: *Escherichia*, *Aerobacter*, *Achromobacter*, *Bacillus*, and *Leuconostoc*.

## FOOD SAFETY IN CANNING

Heat treatment provides another avenue for controlling the viability of microorganisms. The processing conditions required for effective and safe storage of heat-processed foods vary with the type of microorganism being considered. Some cells of microorganisms are rather heat sensitive, many are of medium resistance to heat treatment, while some cells and spores are extremely resistant to heat. However, in all instances, there is a time-temperature relationship which can be observed.

### Temperature-Time Relationship

The temperature-time relationship can be stated simply as the fact that the time required for killing the microorganisms decreases as the temperature of processing increases. To illustrate this, the time required to destroy the spores of flat sour bacteria at 100°C was found to be 20 hours (1200 minutes), 10 hours (600 minutes) at 105°C, and only 1 minute at 135°C. This fact makes it possible to shorten processing times for canning significantly if pressure is available so that higher processing temperatures can be utilized. In addition to saving time, the higher processing temperature results in preparation of higher quality canned foods in several instances.

### Populations

Research also has demonstrated that there is value in avoiding as much contamination as possible prior to the processing of the food, too. When the spore or cell count of microorganisms is high, the time required to insure the death of all of them is longer than when the population is smaller. When the concentration of spores of a thermophile in corn juice was 50,000 per milliliter, the juice had to be heated for 14 minutes to kill all of the spores. In contrast, a population of only 50 spores per milliliter (0.1 percent as great as the first count) required only 8 minutes to kill the spores. Clearly, there is greater safety in home canning if the food is relatively free from contamination prior to the actual heat processing.

The source of the microorganismal contamination also influences the resistance of the cells and spores to heat treatment. The contaminants with origins in soils have been found to be more resistant to heat treatment than those which have been cultured on agar or broth. It is clear that field produce has the potential for containing microorganisms which may be relatively heat resistant.

## Influence of Medium

### pH

The medium in which the microorganisms are being heated has a distinct influence on the rate of destruction. Most spores and cells are at their maximum potential for heat resistance at or near a pH of 7 (approximately neutral in reaction). As the medium becomes more acidic, the resistance to heat decreases more rapidly than it does if the medium is adjusted into the alkaline range, but either side of neutral is more destructive than a pH of 7. This effect is of prime significance in the processing of foods by heat.

Foods for heat processing often are clustered into groups according to their acidity. On this basis, acid foods are often defined as foods with a pH of 4.5 or lower, and low-acid foods are above this pH range. A more precise categorization is suggested (Cameron and Esty) to be:

(1) Low-acid foods (pH 5.3 or higher). These foods would include peas, corn, lima beans, meats, fish, poultry, and milk.

(2) Medium-acid foods (pH 5.3 to 4.5). Spinach, asparagus, beets, and pumpkin are grouped here.

(3) Acid foods (pH 4.5 to 3.7). Tomatoes, pears, and pineapple cluster here, although the newer varieties of tomatoes often are higher in pH than this range.

(4) High-acid foods (pH of 3.7 or lower). Sauerkraut and berries are found in this group.

The type of heat processing required for safety in canned foods is influenced by the acidity of the food being processed. Foods classified as low-acid foods must be processed with the aid of special, pressurized equipment so that the temperatures reached in processing will be high enough to destroy the microorganisms which are present. In other words, meats, fish, poultry, and vegetables need to be processed in pressure canners in the home if they are to be safe. Foods higher in acid, that is, with a lower pH, can be processed safely and in a reasonable length of time in a boiling water bath.

The pH of some foods will change during the heat processing of the food. This is particularly true of the low-acid foods. When the pH is 5.5 or higher, heat processing will result in a drop in the pH of the food; more acidic foods will change pH very little during processing.

### Salt and sugar

Salt in low concentrations in foods being canned will often give some protection to the microorganisms, but this is not always true. Sugar has a similar effect on some microorganisms and spores, but has a destructive effect on others. Salt is protective to *Staphylococcus aureus*, but sugar is destructive. *Escherichia coli* and *Pseudomonas fluorescens* are affected in just the reverse manner, that is, protected by sugar and negatively influenced by sodium chloride.

## Microorganisms and Heat Resistance

### Types of organisms

Most yeasts and their spores are killed quickly during heat processing of foods, and this is also true for most molds. Bacterial spores often are considerably more heat resistant, although there will be much variation between species. For example, the spores of *Bacillus anthracis* are killed in less than 2 minutes at 100°C; spores from *Bacillus subtilis* require 15 to 20 minutes at 100°C; spores from *Clostridium botulinum* require 100 to 330 minutes at 100°C; and flat sour bacteria need more than 17 hours at that temperature. Similarly, there is considerable variation in the length of time required to inactivate enzymes in some foods during heat processing. Many of them will be inactivated at temperatures well below boiling, but peroxidases may fail to be inactivated during normal processing.

The spore-forming bacteria in low-acid and medium-acid foods are the cause of the greatest concern in heat-processing of foods. These microorganisms may be classified as obligate aerobes, as facultative anaerobes, and as obligate anaerobes. The first group, the obligate aerobes, require oxygen for their growth. These microorganisms do not present a problem in home canned foods because the canning methods used result in only very limited amounts of molecular oxygen being present. In addition, the heat treatments required to control the problems of the other two classifications will almost certainly eliminate the obligate aerobes.

The facultative anaerobes may present problems of significance in thermal processing of foods. These are the thermophilic, spore-bearing bacilli which can cause flat-sour spoilage. Flat-sour spoilage is characterized as production of acid with virtually no gas. *Bacillus stearothermophilus* is an example of a microorganism which can cause flat-sour spoilage in low-acid foods. In acid foods, particularly in tomatoes, *Bacillus coagulans* is a facultative anaerobe which is a problem. Others also of concern in acid foods include *Bacillus macerans* and *Bacillus polymyxa*, both of which have been found in spoiled canned fruits. These bacilli are generally not viewed as a problem in low-acid canned foods because they are destroyed when pressure processing is used, as is the traditional approach to canning low-acid items.

The obligate anaerobes are divided into those which are mesophilic and those which are thermophilic. Some of the thermophilic obligate anaerobes, notably *Clostridium thermosaccharolyticum*, produce large quantities of gases, such as carbon dioxide and hydrogen. This production is the end result of their utilization of carbohydrates in the food. This spoilage is referred to as "swell" and often is characterized by an odor of butyric acid or "cheese". These anaerobes present their greatest problem in foods ranging in pH from 4.5 to 5.0. Another thermophilic obligate anaerobe, *Clostridium nigrificans*, produces hydrogen sulfide, leading to a readily identified type of food spoilage. Since the hydrogen sulfide produced by *Cl. nigrificans* is soluble in the liquid in the can, the can does not swell. This is in contrast to the bulging observed with the spoilage caused by *Cl. thermosaccharolyticum*. The interaction of iron with the hydrogen sulfide produced in the former case may result in black food products. This type of spoilage is not seen frequently because the spores are not prevalent, and they are usually destroyed by the heat processing.

The spore-forming mesophilic anaerobe, *Clostridium botulinum*, probably is the most famous of the microorganisms which can cause spoilage in heat processed foods. It is the toxin which is the reason for such great concern, for even very minor amounts of the toxin can be lethal. There are several types of spores which have been identified; types A, B, and E are probably of the greatest significance. Spores of types A and B are more heat resistant than type E; thus it is imperative that adequate processing temperatures be reached to avoid the possible problems these spores can create in low-acid foods. It is not always possible to tell when canned foods may be contaminated with the toxin. As a safety measure, it is essential that all home canned, low-acid foods (i.e., vegetables, poultry, meat, and fish) be boiled actively for 15 minutes prior to even tasting the food. This boiling period will detoxify the toxin which could be present in the food. Dramatic reports of deaths due to botulism are noted from time to time. Proper processing and the precaution of boiling the food before eating eliminate this hazard in the home canned food supply.

From an economic standpoint, there are other mesophilic obligate anaerobes which can cause loss of heat-processed foods if their processing has been inadequate. Potential causes of such spoilage include the failure to kill microorganisms, such as *Clostridium putrificum*, *Clostridium histolyticum*, *Clostridium bifermentans*, and *Clostridium sporogenes*. The greatest problem in low-acid and semi-acid foods (pH 4.5 and above) is caused by *Clostridium sporogenes*, a putrefactive anaerobe. At pH 4.0 to 4.5, the greatest problems with spoilage are caused by butyric anaerobes, such as *Clostridium pasteurianum*.

Spoilage in heat-processed foods with a pH of 4.0 or lower most frequently is caused by non-spore forming bacteria, yeasts, and molds. This is sometimes a problem because these foods ordinarily are processed with more mild heat treatment than those with lower acid content. An example of the problem is provided by *Byssochlamys fulva*, a mold which can cause disintegration of the pectic substances, and consequently of much of the cell structure, in fruits.

**Heat penetration**

Closely related to the destruction of microorganisms by heat treatment is the effectiveness of heat penetration in the food during processing. Heat ordinarily is transmitted throughout the food gradually by a combination of conduction and convection. Conduction in foods is extremely slow compared to the rate of heat transmission by conduction in metals. However, the rate of heat transfer is accelerated by convection currents when the packing of the jar or can facilitates this mode of transmission. An example would be the vertical packing of long green beans or of pickles, as contrasted to the horizontal packing which is found in canned spinach or peaches. Vertical packing facilitates the development of convection currents, and the entire jar and its food mass will more quickly arrive at the necessary processing temperature than will one in which vertical currents are blocked. The need for heat penetration throughout the food mass has been recognized in commercial canning; the result has been the agitation of cans during heat processing to help maintain effective convection currents and insure more uniform heating throughout during commercial processing.

The use of hot pack for foods, rather than cold pack, is another device to help insure that the entire contents of the jar reaches the necessary processing temperature for a sufficiently long period of time. Uniform heat distribution is a greater problem when bulky food items are being processed than when small items, such as peas are being heated. This is because of the greater ease of heat transmission in the smaller, more mobile pieces of food. Viscous foods, such as cream sauces or tomato sauces, require longer heat processing times because the heat transfer is primarily by conduction. Heat moves slowly through pumpkin and other types of viscous foods.

The cooling of canned foods also needs to be considered from the viewpoint of food safety. There are some thermophilic microorganisms which may thrive on foods during the cooling period. To minimize this possible problem, canned foods should be cooled as quickly as is practical within the limitations of the type of container being used. In the home, this simply means removing the jars from the canner and letting them cool on a rack at room temperature with good air flow being permitted between individual jars. This technique is adequate to avoid microbiological problems without jeopardizing the safety of the glass jars.

## FREEZING AND MICROORGANISMS

Freezing is a relatively effective means of slowing growth of microorganisms, although the microorganisms ordinarily are not killed by freezing. The limited growth is the result of often being below the optimum temperature range for growth and reproduction and of the reduced amount of free water available. Enzymatic action may still continue to be relatively active at freezing storage temperatures. This can also cause deteriorative changes in the food. It is for this reason that blanching usually is used as a part of the preparation for freezing vegetables.

Examples of molds which grow at freezer temperatures are *Cladosporium*, *Sporotrichum*, *Penicillium*, and *Monilia*. Examples of cryophilic bacteria are found among *Pseudomonas*, *Achromobacter*, *Alcaligenes*, *Micrococcus*, and *Flavobacterium*. Some growth of these microorganisms has been observed at temperatures ranging as low as -5 to -17°C (23 to 0°F).

Despite these examples of cryophilic microorganisms, freezing is lethal to between 50 and 80 percent of the microorganisms which may be present prior to freezing.

The rate of freezing influences the destructive effect of the freezing process. Slow passage through the temperature range of -1 to -5°C (30 to 23°F) is useful in killing microorganisms, although slow cooling to this temperature range will result in possibly considerable growth of microorganisms before reaching this point of destruction, thus the total number of microorganisms surviving could be greater than would be found with an accelerated rate of freezing. The length of storage time also influences how many viable microorganisms remain. The death rate is very gradual during frozen storage, but, nevertheless, there is some loss of life during prolonged storage, presumably due to starvation of the microorganisms. However, viable microorganisms have been found after several years of frozen storage.

The food medium in which the microorganisms are found also influences the rate of destruction during frozen storage. Foods which are high in water and/or acid may hasten destruction of microorganisms. Conversely, sugar, salt, proteins, and fats are found to be generally protective of microorganisms in frozen foods.

The alternating freezing and thawing of foods containing microorganisms appears to be more destructive to microorganisms than simply keeping the food frozen solidly at all times. However, exceptions to this rule can be found.

## SUMMARY

Spoilage of preserved food during storage may have both health and economic implications. Therefore, attention to correct processing is essential if such problems are to be avoided. Dried foods generally are safe to eat, despite the fact that they will have some viable microorganisms on the surface. The blanching of foods prior to drying, adequate drying, and attention to sanitary methods of storage help to reduce problems with dried foods.

Canned foods, particularly low-acid foods, have the potential for harmful growth of microorganisms during storage. Particular attention needs to be directed toward the adequate processing of low-acid foods (including vegetables, meats, poultry, and fish) to destroy the heat-resistant spores which can ultimately result in the production of dangerous toxins. By boiling low-acid foods for 15 minutes prior to even tasting them, the hazards of possible botulism due to ingestion of viable toxin from *Cl. botulinum* can be avoided.

Frozen foods will often contain some microorganisms and also active enzymes. However, the temperatures utilized in storage are so low that the activity of these detrimental components is reduced significantly. In some instances, frozen storage kills a significant number of the microorganisms which are present. Attention to maintaining adequately cold storage of frozen foods needs to be given in the marketing chain for commercial products, as well as in the home. Blanching is an important technique for helping to inactivate enzymes and kill microorganisms prior to the freezing of vegetables.

## BIBLIOGRAPHY


Anderson, A.W. Significance of yeasts and molds in foods. *Food Tech. 31, No. 2:* 47. 1977.

Cameron, E.J. and J.R. Esty. Comments on the microbiology of spoilage in canned foods. *Food Res. 5:* 549. 1940

Dack, G.M. *Food Poisoning.* University of Chicago Press. Chicago. 3rd ed. 1956.

Frazier, W.C. *Food Microbiology.* McGraw-Hill. New York. 2nd ed. 1967.

Sauer, F. Control of yeasts and molds with preservatives. *Food Tech. 3l, No. 2:* 66. 1977.

Van Arsdel, W.B., et al. *Food Dehydration.* Vol. 2. Avi. Westport, Conn. 1973.

Van Arsdel, W.B., et al. *Quality and Stability of Frozen Foods.* Wiley-Interscience. New York. 1969.

Walker, H.W. Spoilage of foods by yeasts. *Food Tech. 31, No. 2:* 57. 1977.

# Chapter 3.
# Preservation by Drying

## INTRODUCTION

Drying is an intriguing method of preserving foods. Dried foods have the advantage of being easy to transport because of their light weight and of not needing special storage equipment. People undertaking food storage for emergencies are wise to use dried foods because there is no problem of food loss in a crisis due to breaking of containers (as is true of canned foods) nor of thawing due to electrical power failures (as can occur with storage of frozen foods for emergencies). Another situation in which dried foods are particularly appropriate is for persons who are camping or backpacking. Since the food does not have to be maintained in a frozen state or even chilled, refrigeration is not required. People carrying food for any distance will be particularly aware of the significant advantage in using dried foods. By removing most of the water content of a food, the weight of the food is reduced to an absolute minimum. It is the water content in foods which is responsible for much of the weight of foods prior to dehydration. A related advantage is the fact that dried foods, because of their low moisture content, also are reduced greatly in volume, thus making them simpler to store and to pack. All of these advantages accrue as a result of a reduction in moisture content from more than 90 percent in many foods to usually less than 10 percent in the dried product.

## CELL STRUCTURE AND DEHYDRATION

The portions of vegetables and fruits which are consumed as food are comprised primarily of parenchyma cells. These cells may vary significantly in size, but they generally fit the description of being polyhedrons with 14 sides and with fairly rigid cell walls. The actively metabolizing part of a parenchyma cell (the protoplast) is a thin film between the vacuole and the cell wall; the protoplast represents only five percent of the total volume of the cell. The protoplasm is bounded on the cell wall side by a membrane called the plasmalemma, and the membrane on the side of the protoplasm next to the vacuole is the tonoplast. The protoplast and these membranes are composed of protein, lipoproteins, phospholipids, and phytic acid.

The vacuole is a key part of the cell when drying is the mode of food preservation being used because approximately 90 percent of the water in plants is held in this part of the cells. Contents of the vacuole, in addition to water, include salts, organic acids, sugars, amino acids, vitamins, a little oil in droplets, and water-soluble pigments. Evidence of the presence of the water-soluble anthocyanin pigments in the vacuole can be seen when beets are cut, for the cleavage of the vacuole releases this highly visible pigment. Perhaps less obvious, but no less important is the content of flavoring compounds contained in the vacuole. The amino acids and sugars in the vacuole are also of interest in the drying process because of their potential for combination and ensuing color changes due to the browning reaction which can occur if enzymes from the cytoplasm should happen to enter the vacuole.

The third segment of cell structure is the cell wall. Included in the components are the primary wall, the middle lamella, the plasmodesmata, and surface materials. The primary wall encases the individual parenchyma cell, lying adjacent to the plasmalemma. Constituents of the primary wall include pectic substances, cellulose, hemicellulose, and non-cellulose polysaccharides. Between cells lies the middle lamella, a structure serving to

## PARENCHYMA CELL

**Vacuole**

**Protoplast**

- Tonoplast
- Protoplasm
- Nucleus
- Cytoplasm
  - Mitochondria
  - Ribosomes
  - Lysosomes
  - Microsomes
  - Endoplasmic reticulum
  - Golgi bodies
  - Chloroplasts
  - Chromoplasts
  - Leucoplasts
- Plasmalemma

**Cell Wall**

- Primary wall
- Middle lamella
- Plasmadesmata

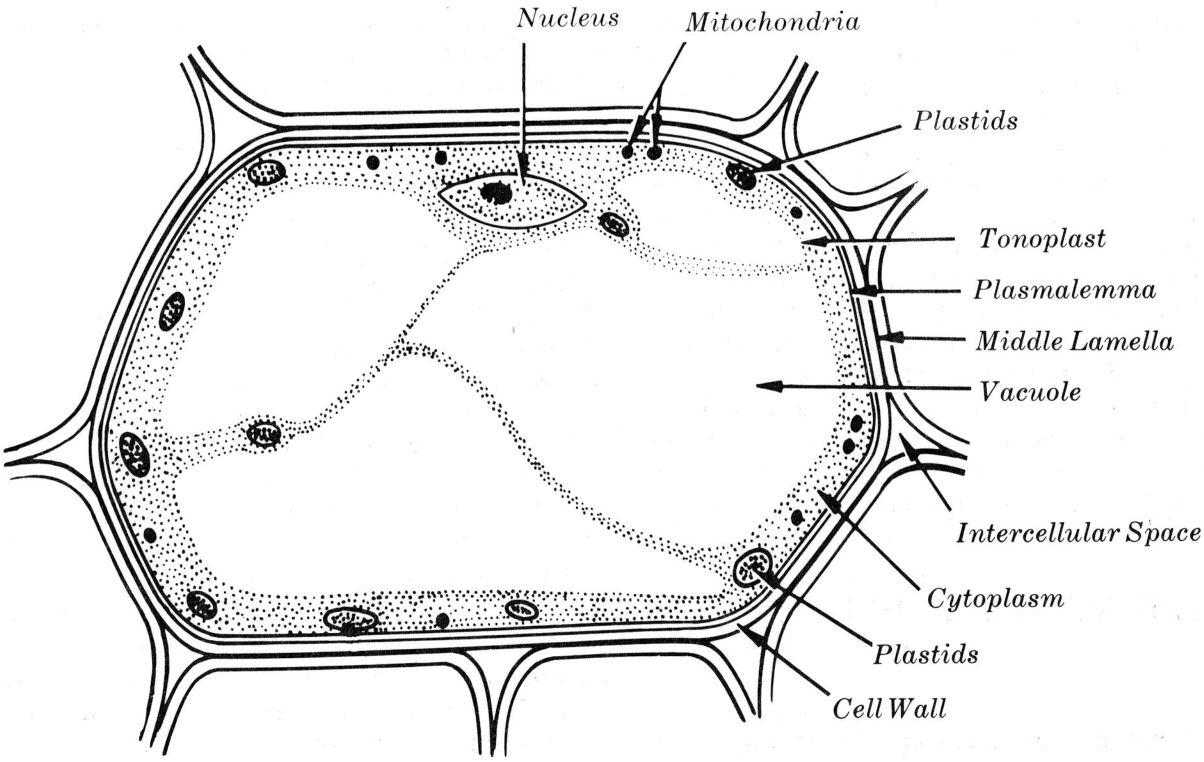

## Chemical Components of Parenchyma Cell

### Vacuole

- Water
- Salts
- Amino acids
- Vitamins
- Oil droplets
- Water-soluble pigments
  - Anthocyanin
  - Anthoxanthin
- Organic acids
- Sugars
- Flavoring compounds

### Protoplast

- Protein
- Lipoproteins
- Phospholipids
- Water-insoluble pigments
  - Carotenoids
  - Chlorophyll
- Phytic acid
- Starch
- Enzymes
- Vitamins
- Minerals
- Lipids

### Cell Wall

- Cellulose
- Pectic substances
- Hemicelluloses
- Proteins
- Other polysaccharides

cement cells together; composition of the middle lamella includes primarily pectic substances, some non-cellulose polysaccharides, and some calcium and magnesium salts.

## The Cell Wall

The foundation of the cell wall of parenchyma cells is cellulose arranged in a somewhat organized network of microfibrils. The pectic substances, non-cellulose polysaccharides, and hemicelluloses are deposited on this surface and in the cavities presented within this cellulose network.

Some vegetables may have additional tissue deposited on this basic structure to produce the lignified structures sometimes seen in mature vegetables. Celery provides an example of the development of special tissues deposited on the basic microfibril network of the primary wall. This specialized tissue is composed of collenchyma cells, which give the structural support noted in celery.

The middle lamella is the cementing material of plants. The softening observed during the cooking of vegetables and fruits, as well as in the ripening process, appears to be due to changes in the pectic substances comprising much of the structure of the middle lamella. Proteins and additional polysaccharides are the other constituents of the middle lamella. When vegetables are cooked, the calcium and magnesium divalent cations present have the potential for entering the interior of the cell or adhering to the cell wall. The phytic acid in the vacuole competes actively with the pectic substances in the cell wall for these cations. If the minerals are drawn into the vacuole, the cells will be held together less tightly, and the texture will soften. Conversely, if the calcium and magnesium cations are held by the pectic substances, softening will be retarded greatly. During drying, these various cell wall components will become dehydrated, accounting for part of the loss of weight and water which occurs. Fortunately, these polysaccharides are rehydrated readily to essentially their original volume, making the use of dried foods a practical matter.

The remaining segment of cell wall components is the plasmodesmata. This type of structure is described as being strands of cytoplasm which interconnect the cytoplasm of adjoining cells by passing through the cell walls into the protoplasts.

Also considered to be a structural feature of plant tissues is intercellular space. Intercellular spaces provide a route for interconnecting the internal atmosphere of the vegetable or fruit.These spaces generally are found where the facets of the cells meet.

## The Protoplast

The nucleus, cytoplasm, and membranes surrounding the protoplast comprise the total portion of the cell known as the protoplast. The nucleus is unique because of its content of genetic information. This is where DNA (deoxyribonucleic acid) and other proteins, including enzymes, are found. Despite its significance for growth, the nucleus contributes only about five percent of the total protein content found in the protoplast.

The cytoplasm of the protoplast contains the mitochondria, ribosomes, lysosomes, microsomes, endoplasmic reticulum, and Golgi bodies as key components. Vitamins, minerals, and enzymes are concentrated largely in the cytoplasm. The trace minerals contained in the cytoplasm may be freed from the coenzyme form in which they are bound prior to drying. When this happens, iron, cobalt, calcium, magnesium, molybdenum, or other trace minerals may be enabled, via the process of drying, to interact with compounds with which they come in contact. Oxidation of vitamins or detrimental color changes are possible in dried foods as a result of the liberation of these mineral ions.

The chloroplasts in the cytoplasm of the protoplasts are significant as the site of photosynthesis. The chlorophyll of plants is contained in the chloroplasts. This pigment can fairly readily undergo a chemical change in which the magnesium atom is replaced by hydrogen, with an accompanying color change from green to olive green.

Chromoplasts also are contained in the cytoplasm of the protoplasts. In these structures are found the carotenoid pigments. These pigments are not water soluble.

An important part of cytoplasm is the structure called the leucoplast. Within the leucoplasts, the starch granules are deposited, serving as a reserve source of carbohydrate in the cell. The starch deposited in the leucoplasts gelatinizes when vegetables are blanched or cooked prior to drying. When the vegetables are dried following the heating period, the water must pass through the resulting starch gel. Thus, the amount of gelatinized starch found in a vegetable will influence the rate and the temperatures which may be used for dehydrating the vegetable. Indeed, there will even be differences noted between different varieties of potatoes.

Most of the protein contained within a plant cell is found in the protoplast, and these proteins are primarily enzymes. Some protein is stored as a reserve supply in the aleurone grains in the cytoplasm. During blanching or cooking prior to drying, the proteins are denatured. This results in textural changes, particularly in membranes. Proteins also bind water, a fact which is of significance in drying. If severe procedures are used in drying, some of the bound water may be removed, resulting in irreversible changes in the dehydrated food.

Lipids in the protoplast of cells are of special interest because of the influence they may have on the flavor of dehydrated foods. Oil droplets are composed mostly of esters of fatty acids and glycerol. In the presence of lipoxidase, hydrolysis of these esters to the component fatty acids and glycerol can occur, with oxidation of the unsaturated fatty acids. The result is a detrimental change of flavor. Vitamin A activity also can be reduced due to oxidation. Such changes can occur in dehydrated vegetables even when enzymes have been inactivated by heat prior to drying, but blanching is beneficial in reducing this problem.

Although the cell membranes of the protoplast are comprised mainly of the plasmalemma and the tonoplast, other components also should be noted. These include the endoplasmic reticulum and the membranes surrounding the nucleus and the mitochondria.

## The Vacuole

The vacuole is of particular interest when one considers palatability of foods because many of the important compounds for flavor are contained in this portion of the cell. When the contents of the vacuole come in contact with the enzymes from the protoplast, as can occur in drying, significant changes may be the result.

If desirable flavors are produced by enzyme action, it is imperative to avoid destruction of the enzyme during processing. In onions, garlic, and horseradish, this is exactly the case. Preparation for drying these foods should not include blanching because blanching inactivates the enzyme (alliinase) which is needed to convert derivatives of cysteine sulfoxide into the allium compounds that contribute the characteristic odor and flavor. Similarly, enzymatic action is needed to develop the desired flavor in cabbage and other vegetables with related flavoring components. In these, glucosides are converted to isothiocyanate derivatives by enzyme action when the vacuole and the protoplast come in direct contact.

Not all enzymatic changes result in pleasing flavors. An example of a detrimental change can be found in green beans. For this reason, many vegetables are blanched prior to drying.

Color changes resulting from enzyme action also can be noted. For example, polyphenol oxidase oxidizes phenolic compounds to melanins which are observed as darkened regions in the food. Tyrosine is an amino acid intimately involved with the darkening in some plant foods.

The vacuole contributes to the texture of a vegetable or fruit because of the semipermeability of the tonoplast. This membrane is permeable to water, but not to salts and sugar within the vacuole. Consequently, osmotic pressure can develop, with the direction of the water being dictated by the external conditions surrounding the cell. Of interest in drying is the fact that salt brine would draw water out of the cell vacuole, thus helping to shrink the vacuole and causing the cytoplasm to shrink away from the cell wall. Perhaps more applicable is the fact that, when the cell is killed by blanching, the cell wall becomes much more permeable to both the solute and the water. This change facilitates water removal during drying.

# TYPES OF DRYING

As can be deduced from the fact that drying was the earliest form of food preservation, this is the preservation technique requiring the least equipment for successfully making food safe for long-term storage. In fact, drying often was done in the past with no equipment. Dehydration was accomplished simply by spreading pieces of the food on the ground in the sun or stringing it between trees and leaving it exposed until dried sufficiently for storage. This sun-drying technique was used for centuries despite the fact that the food was subjected to the hazards of insects and inclement weather which could make the food unpalatable and even unsafe to eat. If rain or cloudy, cool weather occurred while food was being dried, the drying food often would be spoiled and unfit for consumption.

Today food may be dried in the home utilizing sun drying (with appropriate refinements on the original, primitive method), oven drying, or drying in a dehydrator. Any of these methods is simple to do, and the equipment required ranges from no extra equipment for oven drying to purchase or construction of a simple dehydrator. The storage containers for the dried foods also are simple and inexpensive.

## Principles

### Sun drying

Sun drying is a technique of drying which is of particular interest today because of the fact that no energy is required for the drying process. Instead, the energy is provided by the sun. For this to be a practical method, however, it is desirable to have hot days, preferably at least 100°F, and night temperatures which remain within 20° of the daytime temperatures. Additionally, a low humidity is essential. Molding or souring of the food will occur if drying is too slow.

A disadvantage of sun drying is that it is much slower than the other two methods of drying. Under optimum conditions, as much as three or four days are required to dry foods to a satisfactory moisture level. With more moisture or cooler temperatures, even longer periods may be required. Also, rain during the drying period can cause complete loss of the food being dried.

### Oven drying

Drying in the oven is a very convenient approach to drying, particularly when the climate is not appropriate for sun drying and when a person is experimenting with the use of drying as a food preservation technique. Drying in the oven requires only an appropriate screen or net for holding the food as it dries. A conventional oven with a thermo-

*Figure 3.1. Raisins traditionally are sun dried.*

*Figure 3.2. Oven drying can be done simply by securing a piece of netting to the regular oven rack to provide the necessary support for the food without impeding the circulation of air currents in the oven.*

static setting of 140°F is suitable for oven drying of fruits, vegetables, fruit leathers, and meats. A microwave oven is not well suited to drying of foods. The low moisture content of the foods as they approach the desired final state is not compatible with the design principles of the unit, and the magnetron tube may be damaged.

**Dehydrator**

Dehydrators for drying foods can be purchased or constructed rather easily at home. Their rather unsophisticated design allows them to be built at home at a fairly low cost; the commercial products are distinctly more expensive than the homemade dehydrators. The dehydrator basically is a series of framed screens which fit onto guides housed in a chamber containing a low-powered fan and heater. The heater maintains the interior of the chamber at 140°F (60°C) throughout the drying period. The fan is helpful in promoting a uniform temperature throughout the chamber; it also aids in removing moisture from the drying area.

*Figure 3.3. A dehydrator with a fan for promoting uniform heating of the food and removal of moisture. Note the shelf guides for the trays.*

A natural-draft dehydrator is a variation of the dehydrators outlined in the previous paragraph. Both the heat source and the fan are omitted from this simple dehydrator. Instead, the trays are staggered in the compartment (one shelf extending all the way to the back of the compartment, while the next shelf reaches from the front only part way to the back), and this alternating pattern of placement is utilized to facilitate heat flow throughout the compartment so that a fan is not required. The heat source is provided externally by a vertical heater, which may be a gas wall heater or a portable electric heater.

The various types of dehydrators are useful in drying large amounts of food. They can be used regardless of the weather. Furthermore, it is not necessary to coordinate drying with normal use of the oven for baking and meal preparation. This makes dehydrators particularly attractive to persons who dry foods frequently and/or in large quantities.

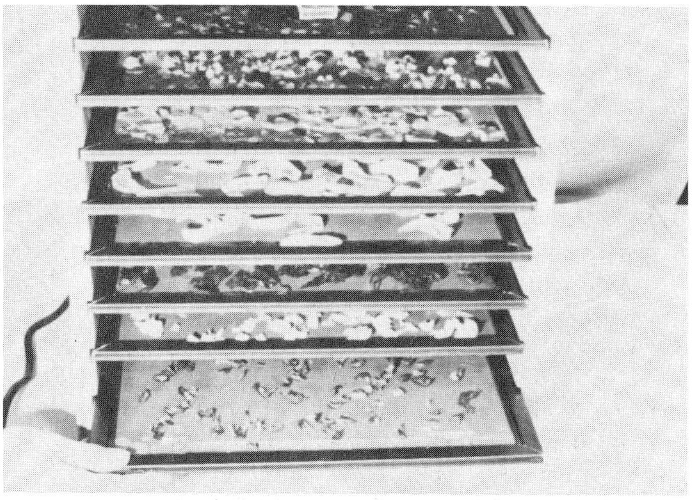

*Figure 3.4. Although dehydrators involve some initial expense, they provide a convenient and effective means of drying relatively large quantities of food at the same time without fear of insects or other external contaminants.*

The disadvantages of dehydrators are the initial investment required and the need for storing the equipment when it is not being used. For people with large amounts of food to dry and limited time for accomplishing the drying, a dehydrator provides a practical solution, however. A word of caution also should be mentioned when considering natural-draft dehydrators for they do have the disadvantage of presenting a possible fire hazard from the external heat source.

## EQUIPMENT FOR DRYING

### Sun Drying

Although sun drying can be done simply by placing food in the sun, it is far more desirable to use a rack on legs and equipped with a fixed screen in the bottom and an upper screen which can be removed to allow the food to be placed in the drying compartment. This upper screen should be designed so that it can be attached to the frame tightly. This design provides good air circulation and keeps insects from the food. Light-weight units make it easy to move the food into the house at night, thus reducing the possibility of having food spoil during drying. These compartments can be constructed or bought ready to use.

### Oven Drying

The original oven racks can be used very satisfactorily for drying. It is necessary to place a portable screen or to fasten some netting over the rack to keep the food from falling as it contracts during drying. A piece of aluminum foil in the bottom of the oven is useful in protecting the oven floor from any dripping which might occur during the first part of the drying period. This is particularly useful when marinated meats or juice-coated fruits are being dried.

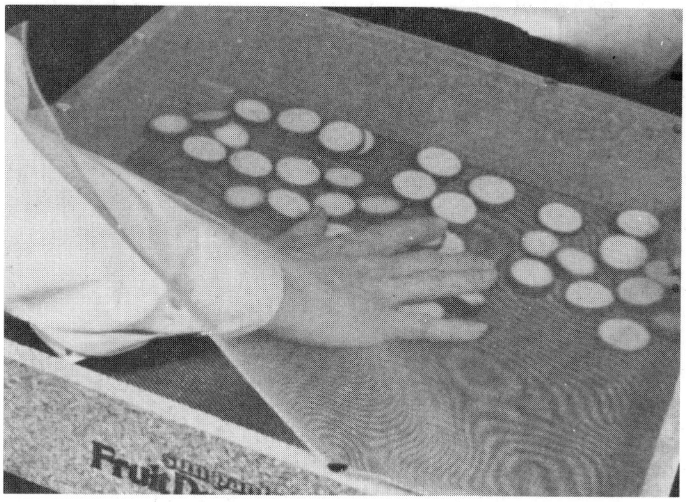

*Figure 3.5. A well-designed tray for sun drying will consist of a compartment on legs. The floor of the compartment is a screen permanently attached to the frame; the upper screen can be folded back for convenience in loading and unloading the food.*

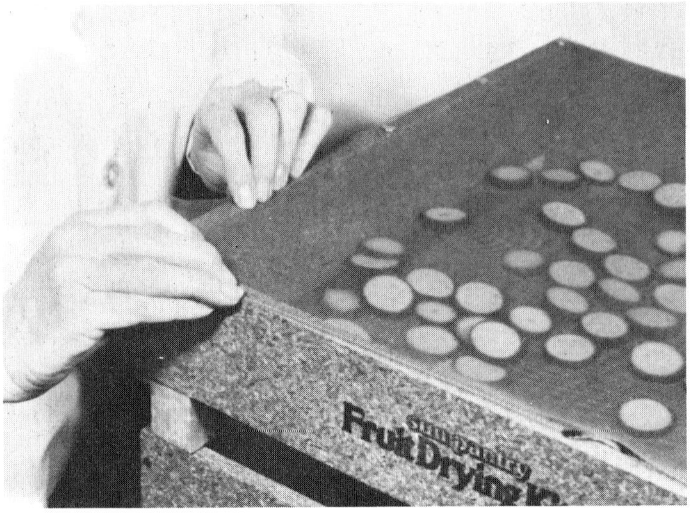


*Figure 3.6. The upper screen needs to be able to be fastened securely to prevent entry of insects or animals while the food is being dried in the sun.*

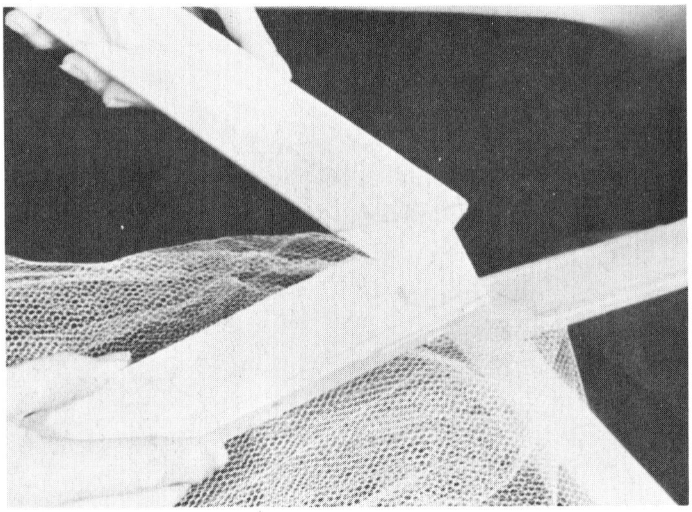

*Figure 3.7. Racks for oven drying can be made inexpensively from unfinished stretcher bars and netting.*

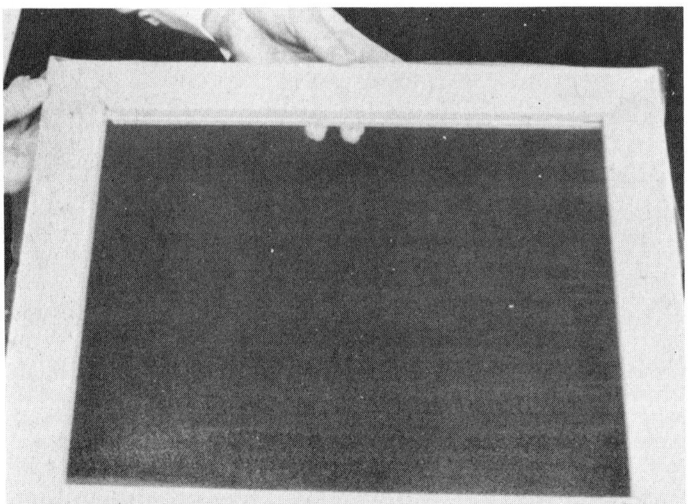

*Figure 3.8. The assembled screen frame and netting provides a rigid screen which is particularly well suited to oven drying.*

*Figure 3.9. These home-fashioned trays can be made the appropriate size to fit a specific oven, thus enabling optimal use of the oven capacity for drying.*

*Figure 3.10. If netting is being used directly on the oven rack, it must be fastened very securely to the rack so that the weight of the food will not pull any part of the netting loose enough to allow the food to slide together or to drop from the rack.*

### Dehydrator

A dehydrator is the only piece of equipment needed when this method of drying is selected. If the heat source in the dehydrator is light bulbs, it is important to be sure that bulbs of the correct wattage are available. The screens should be made of stainless steel hardware cloth or fiber glass screen material. A thermometer for checking the temperature of the dehydrator (even if thermostat is present) allows a simple check that the dehydrator is not exceeding 140°F (60°C).

## FOODS FOR DRYING

### Selection

One of the motivations for preserving foods is to save money on food costs. This well may be a reason why people decide to dry food. Money will be saved if foods for drying are purchased at a time of year when they are plentiful and, consequently, relatively less expensive. If a person has a garden, drying may be a very convenient way of preserving at least a part of the harvest, and the food is available merely for the cost of the seeds, fertilizer, water, and labor. Drying may provide an economy for families who might buy more potatoes, onions, green peppers, apples, or parsley than will be consumed while the quality is high. Part of a bunch of parsley is definitely inexpensive to dry if only a fraction of the purchased bunch will be used in regular food preparation.

Another reason for drying food may be the desire to have special foods available which are of higher quality than commercial counterparts. For example, one might greatly prefer a favorite jerky recipe to the commercial beef versions available. There also can be greater variety of dried foods when they are prepared at home. In addition, personal satisfaction is gained from creating these unique products.

Only foods of high quality should be dried. If blemishes are noted, they should be removed before drying is begun. Soft and over-ripe foods will produce disappointing results in the finished product.

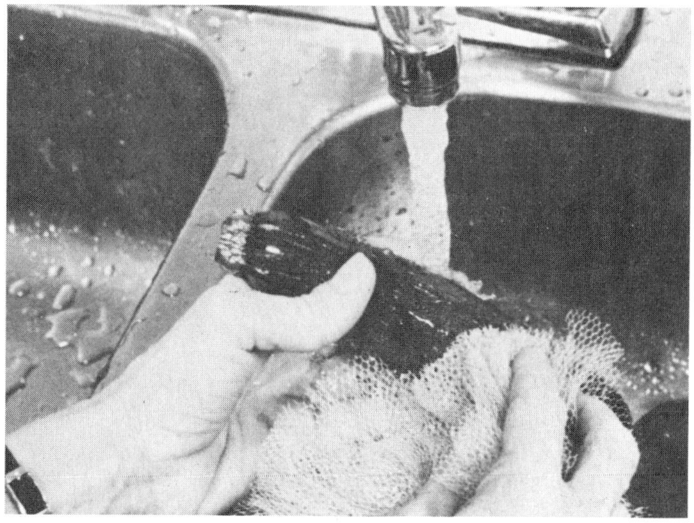

*Figure 3.11. One means of helping to make dried foods safe is to remove all soil (and the microorganisms it contains) prior to drying. Vigorous scrubbing with a vegetable brush, coarse netting, or other abrasive material is helpful in removing dirt that tends to cling. Running water helps to carry away the loosened dirt.*

## Preparation of Foods

Prior to drying, all foods need to be impeccably clean. A vegetable brush or a piece of netting can be used effectively to help loosen any particles of dirt which may be clinging to fresh produce. Mushrooms are particularly prone to collecting dirt. These need to be rubbed gently with the fingers while they are held under a stream of running water so that all of the dirt will be removed. The stem area of tomatoes needs to be cut out with a paring knife. Pomes, such as apples, need to be cored. The specific preparation of selected fruits and vegetables is outlined in Tables 3.1 and 3.2.

*Figure 3.12. Mushrooms often have considerable dirt clinging to them. Thorough washing prior to drying is imperative, both for sanitation and texture.*

In addition to the removal of tough areas of the food, such as the core, it is essential that each item be examined carefully for any suggestion of spoilage or soft spots. All such areas should be cut out before the food is dried. Failure to do this increases the likelihood of lower quality and possible spoilage.

Some foods need to have their skins removed before they are dried. A good example of this is provided by winter squash. Almost all foods also will need to be cut into slices of the appropriate size for easy drying. The thickness may range from as little as ⅛ inch to as thick as ¾ inch, depending upon the type of product being dried. However, it is recommended that uniform pieces be placed on a single tray for convenience in the actual drying process. Okra, onions, potatoes, and carrots are examples suitable for preparation in very thin slices. In contrast, tomatoes for stewing can be cut into ¾-inch wedges. For most foods, a thickness of ¼ inch or a little more is desirable.

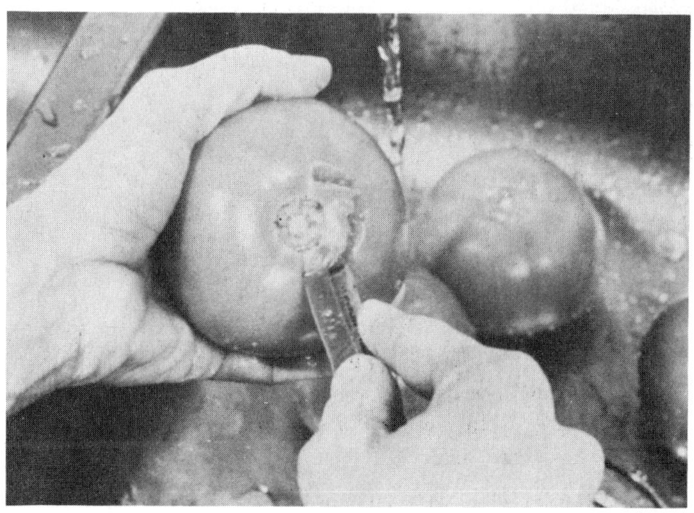

*Figure 3.13. The stem end of the tomato is removed as efficiently as possible in preparation for drying.*

**Table 3.1. Home drying of vegetables[1]**

For oven and portable dehydrators, set temperatures at 140° F.
Sun drying requires temperatures of 98° to 100 ° F.

| Vegetable | Preparation | Blanching | | Drying | |
|---|---|---|---|---|---|
| | | Method | Time | Method | Time |
| | | | minutes | | hours |
| Artichoke, globe | Cut hearts into 1/8 inch strips. | Heat in boiling solution (3/4 cup water, 1tbs. lemon juice.) | 6 - 8 | dehydrator<br>oven<br>sun | 12 -16<br>4 - 7<br>10 -12 |
| Asparagus | Wash thoroughly. Halve large tips. | Steam<br>Water | 4 - 5<br>3½- 4½ | dehydrator<br>oven<br>sun | 6 -10<br>3 - 4<br>8 -10 |
| Beans, green | Wash thoroughly. Cut in short pieces or lengthwise. | Steam<br>Water | 2 - 2½<br>2 | dehydrator<br>oven<br>sun | 8 -14<br>3 - 6<br>8 |
| Beets | Cook as usual. Cool; peel. Cut into shoestring strips 1/8 inch thick. | Already cooked; no further blanching required. | — | dehydrator<br>oven<br>sun | 10 -12<br>3 - 5<br>8 -10 |
| Broccoli | Trim, cut as for serving. Wash thoroughly. Quarter stalks lengthwise. | Steam<br>Water* | 3 - 3½<br>2 | dehydrator<br>oven<br>sun | 12 -15<br>3 - 4½<br>8 -10 |
| Brussels sprouts | Cut in half lengthwise through stem. | Steam<br>Water | 6 - 7<br>4½- 5½ | dehydrator<br>oven<br>sun | 12 -18<br>4 - 5<br>9 -11 |
| Cabbage[2] | Remove outer leaves; quarter and core. Cut into strips 1/8 inch thick. | Steam until wilted<br>Water | 2½- 3<br>1½- 2 | dehydrator<br>oven<br>sun | 10 -12<br>1 - 3<br>6 - 7 |
| Carrots | Use only crisp, tender carrots. Wash thoroughly. Cut off roots and tops; preferably peel, cut in slices or strips 1/8 inch thick. | Steam<br>Water | 3 - 3½<br>3½ | dehydrator<br>oven<br>sun | 10 -12<br>3½- 5<br>8 |
| Cauliflower | Prepare as for serving. | Steam<br>Water* | 4 - 5<br>3 - 4 | dehydrator<br>oven<br>sun | 12 -15<br>4 - 6<br>8 -11 |
| Celery | Trim stalks. Wash stalks and leaves thoroughly. Slice stalks. | Steam<br>Water | 2<br>2 | dehydrator<br>oven<br>sun | 10 -16<br>3 - 4<br>8 |
| Corn on the cob | Husk, trim. | Steam until milk does not exude from kernel when cut<br>Water | 2 - 2½<br>1½ | dehydrator<br>oven<br>sun | 12 -15<br>4 - 6<br>8 |
| Corn, cut | Prepare in the same manner as corn on the cob, except cut the kernels from the cob after blanching. | | | dehydrator<br>oven<br>sun | 6 -10<br>2 - 3<br>6 |
| Egg plant | Use the same directions as for summer squash. | Steam<br>Water | 3½<br>3 | dehydrator<br>oven<br>sun | 12 -14<br>3½- 5<br>6 - 8 |
| Horseradish | Wash; remove small rootlets and stubs. Peel or scrape roots. Grate or cut into 1/8-inch slices. | None | — | dehydrator<br>oven<br>sun | 4 -10<br>3 - 4<br>7 -10 |

*Continued.*

[1]From M.W. Miller, et al. *Drying Foods at Home.* Home and Garden Bulletin No. 217. U.S. Dept. Agriculture. Washington, D.C. P. 15. 1977.

[2]Blanching of cabbage and related vegetables is optional.

**Table 3.1. Continued**

| Vegetable | Preparation | Blanching | | Drying | |
|---|---|---|---|---|---|
| | | Method | Time | Method | Time |
| | | | minutes | | hours |
| Mushrooms (WARNING, see below)† | Scrub thoroughly. Discard any tough, woody stalks. Cut tender stalks into short sections. Do not peel small mushrooms or "buttons." Peel large mushrooms, slice. | None | — | dehydrator..........<br>oven....................<br>sun...................... | 8 -10<br>3 - 5<br>6 - 8 |
| Okra | Wash, trim, slice crosswise in 1/8 - 1/4 inch disks. | None | — | dehydrator..........<br>oven....................<br>sun...................... | 8 -10<br>4 - 6<br>8 -11 |
| Onions | Wash, remove outer "paper shells." Remove tops and root ends, slice 1/8 - 1/4 inch thick. | None | — | dehydrator..........<br>oven....................<br>sun...................... | 10 -20<br>3 - 6<br>8 -11 |
| Parsley | Wash thoroughly. Separate clusters. Discard long or tough stems. | None | — | dehydrator..........<br>oven....................<br>sun...................... | 1 - 2<br>2 - 4<br>6 - 8 |
| Peas | Shell. | Steam......................<br>Water........................ | 3<br>2 | dehydrator..........<br>oven....................<br>sun...................... | 8 -10<br>3<br>6 - 8 |
| Peppers and pimientos | Wash, stem, core. Remove "partitions." Cut into disks about 3/8 by 3/8 inch. | None | — | dehydrator..........<br>oven....................<br>sun...................... | 8 -12<br>2½- 5<br>6 - 8 |
| Potatoes | Wash, peel. Cut into shoe-string strips 1/4 inch thick, or cut in slices 1/8 inch thick. | Steam......................<br>Water........................ | 6 - 8<br>5 - 6 | dehydrator..........<br>oven....................<br>sun...................... | 8 -12<br>4 - 6<br>8 -11 |
| Spinach and other greens (kale, chard, mustard) | Trim, wash very thoroughly. | Steam until thoroughly wilted.....<br>Water........................ | 2 - 2½<br>1½ | dehydrator..........<br>oven....................<br>sun...................... | 8 -10<br>2½- 3½<br>6 - 8 |
| Squash: | | | | | |
| Banana | Wash, peel, slice in strips about 1/4 inch thick. | Steam......................<br>Water........................ | 2½- 3<br>1 | dehydrator..........<br>oven....................<br>sun...................... | 10 -16<br>4 - 5<br>6 - 8 |
| Hubbard | Cut or break into pieces. Remove seeds and cavity pulp. Cut into 1 inch wide strips. Peel rind. Cut strips crosswise into pieces about 1/8 inch thick. | Steam......................<br>Water........................ | 2½- 3<br>1 | dehydrator..........<br>oven....................<br>sun...................... | 10 -16<br>4 - 5<br>6 - 8 |
| Summer | Wash, trim, cut into 1/4 inch slices. | Steam......................<br>Water........................ | 2½- 3<br>1½ | dehydrator..........<br>oven....................<br>sun...................... | 10 -12<br>4 - 6<br>6 - 8 |
| Tomatoes, for stewing | Steam or dip in boiling water to loosen skins. Chill in cold water. Peel. Cut into sections about 3/4 inch wide, or slice. Cut small pear or plum tomatoes in half. | Steam......................<br>Water........................ | 3<br>1 | dehydrator..........<br>oven....................<br>sun...................... | 10 -18<br>6 - 8<br>8 -10 |

* Preferred method.

† WARNING: The toxins of poisonous varieties of mushrooms are **not** destroyed by drying or by cooking. Only an expert can differentiate between poisonous and edible varieties.

## Table 3.2. Home drying of fruits[1]

| Fruit | Preparation | Treat before drying. Choose one of the following three methods. | | | Drying times | | Test for dryness |
|---|---|---|---|---|---|---|---|
| | | Sulfur | Steam blanch | Water blanch | Sun drying | Dehydrator | (Cool before testing) |
| Apples | Peel and core, cut into slices or rings about 1/8 thick. | 45 minutes | 5 minutes, depending on texture | —— | 3-4 days | 6-12 hours | Soft, pliable, no moist area in center when cut |
| Apricots | Pit and halve for steam blanch or sulfuring. Leave whole for water blanch. Pit and halve after blanch. | 2 hours | 3-4 minutes | 4-5 minutes | 2-3 days | 24-36 hours | Same as for apples |
| Figs | In dry, warm, sunny climates, it is preferable to partly dry on the tree. Figs normally drop from the tree when 2/3 dry. In coastal areas, pick fruit when ripe. | No treatment necessary | | | 4-5 days | 12-20 hours | Flesh pliable, slightly sticky, but not wet |
| Grapes: Muscat, Tokay, or any seedless grape | Leave whole. Grapes dry in less time if dipped in lye 10 seconds.* Or blanch 1/2 to 1 minute. | No treatment necessary | | | 3-5 days | 12-20 hours | Raisin-like texture, no moist center |
| Nectarines and peaches | When sulfuring, pit and halve; if desired, remove skins. For steam and water blanching, leave whole, then pit and halve. | 2-3 hours | 8 minutes | 8 minutes | 3-5 days | 36-48 hours | Same as for apples |
| Pears | Cut in half and core. Peeling preferred. | 5 hours | 6 minutes (peeled, will be soft) | —— | 5 days | 24-36 hours | Same as for apples |
| Persimmons | Use firm fruit when using the long, softer variety, and use riper fruit when using the round, drier variety. Peel and slice with stainless steel knife. | No treatment necessary | | | 5-6 days | 18-24 hours | Light to medium brown, tender, but not sticky |
| Prunes | For sun-drying, dip in boiling lye solution or blanch in boiling water or steam for 1 to 1-1/2 minutes, to "check" skins. For oven-drying, rinse in hot tap-water. Leave whole. If sulfuring, pit and halve. | No treatment necessary. However, sulfuring for one hour will produce good flavor. | | | 4-5 days (3-4 days when using lye.) | 24-36 hours | Leathery; pit should not slip when squeezed if prune not cut. |

[1]From M.W. Miller, et al. *Drying Foods at Home.* Home and Garden Bulletin No. 217. U.S. Dept. Agriculture. Washington, D.C. P. 15. 1977.

*Figure 3.14. Celery, like most other vegetables, is cut into slices approximately ¼ inch thick to facilitate drying.*

*Figure 3.15. A variety of devices suitable for blanching in boiling water: (left to right) colander, deep-fat fryer basket, cheesecloth, and a strainer.*

## DRYING PROCEDURES

### Vegetables

Deterioration of color and flavor can occur if many of the vegetables are not heat-treated before drying. Heat-treatment is done by boiling the prepared vegetable very briefly in water or by steaming it. Either of these two techniques is termed *blanching*; the former is water blanching, and the latter is called steam blanching. The purpose of this procedure is to inactivate enzymes in the food so that the reactions they otherwise would catalyze will not take place. For most vegetables, quality of the finished product is enhanced by this brief procedure.

The simplest way of blanching is to bring enough water to a boil to cover the food being blanched. No more than two layers of vegetable should be immersed in the water at one time. When vegetables are blanched loose in the boiling water, it is difficult to get all of them removed without undue delay. Consequently, some of the vegetable will likely be overcooked.

There are various devices which may be used conveniently to water blanch vegetables for drying. A perforated basket, such as one used for deep-fat frying, holds a reasonable quantity of food and is easy to use. Its secure handle makes it easy to remove from the boiling water without risk of burns. Other alternatives which can be used, although less conveniently, include a colander, a sieve, or simply a generous piece of cheesecloth or netting.

Similar devices can be used for steam blanching. For example, vegetables to be blanched can be cut into the desired thickness, placed on a large piece of cheesecloth, and then held securely under the dome of a curved lid by grasping the cheesecloth firmly atop the lid. This assembly then is placed securely over its lower container in which water is boiling actively to generate the steam needed

*Figure 3.16. To use a roaster lid and cheesecloth for blanching, the food is distributed on the cheesecloth to cover an area approximating that covered by the roaster lid.*

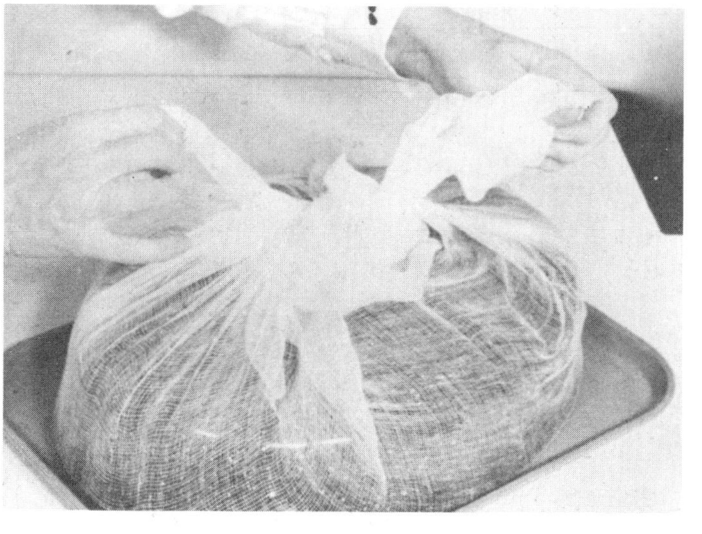

*Figure 3.17. The lid is placed over the food, and the ends of the cheesecloth are fastened together on top of the cover.*

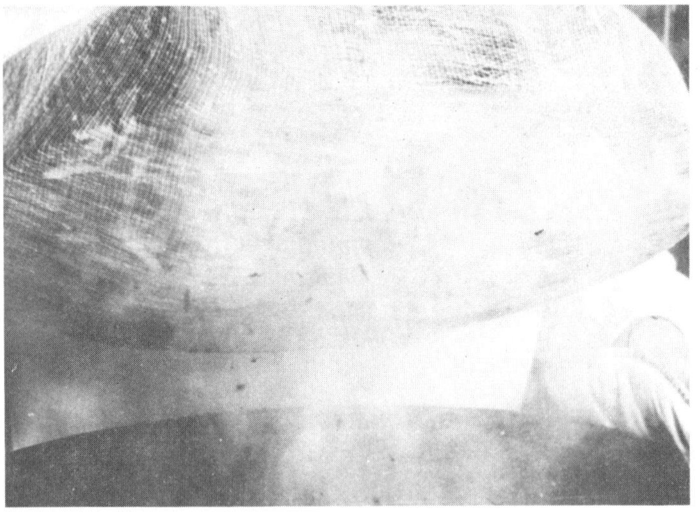

*Figure 3.18. Now the food-cheesecloth-lid assembly is placed over water which is boiling actively in the bottom of the roaster.*

for steam blanching. This technique allows excellent circulation of steam throughout the food. A colander with legs or a steamer rack also can be used to hold the food in the steam above actively boiling water in a covered pan. It is even possible to perforate an aluminum foil plate to improvise a rack for steam blanching. Steam blanching retains more of the water-soluble nutrients than does water blanching.

Immediately after blanching, the food is plunged into ice water to halt the cooking process. Then the food is spread out on toweling to help dry the surface before it is placed in the dryer. Tomatoes are peeled after cooling in ice water. The brief blanching period and the sharp temperature contrast afforded by the ice water make it a very simple matter to remove the loosened skin.

Some vegetables do not need to be subjected to blanching. In fact, their quality will be better if they are not blanched. Examples of foods which should not be blanched prior to drying include horseradish, mushrooms, okra, onions, parsley, cabbage, and other vegetables with cabbage-like flavors.

*Figure 3.19. A colander suspended above boiling water provides a convenient method for steam blanching.*

*Figure 3.20. Steamer racks are particularly good devices for holding vegetables in steam for blanching. Note that the cover to the vessel containing the boiling water must always be used to trap the steam in the pan. Without doing this, the steam escapes too readily, and blanching is unnecessarily slow.*

*Figure 3.21. Another inexpensive solution to providing a rack for blanching is provided by puncturing numerous holes in an aluminum foil pie plate, using a punch can opener to make the perforations.*

*Figure 3.22. The pie plate must fit the top of the pan so that the food will be held in the steam, rather than falling into the boiling water below.*

*Figure 3.23. When green beans or other vegetables are being blanched in preparation for drying, the timing needs to be done accurately.*

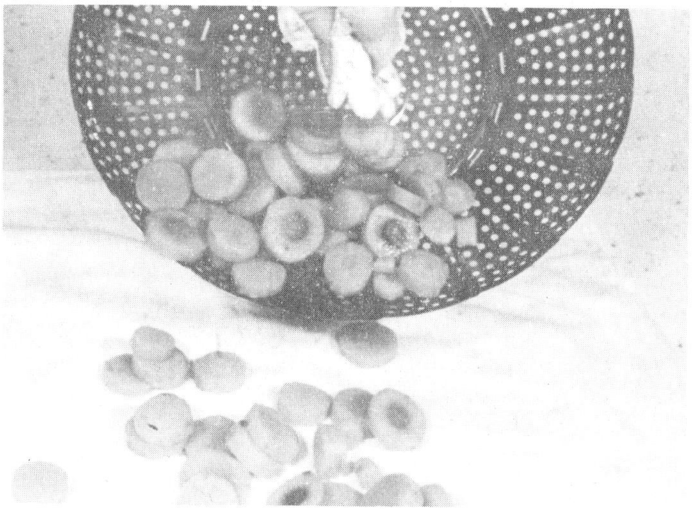

*Figure 3.24. By spreading the blanched carrots on an absorbent towel, the excess moisture will be removed, thus speeding drying. Blotting with the towel further aids the subsequent drying operation.*

*Figure 3.25. The large surface area resulting from the need to cut thin slices for drying represents a significant cause of loss of water-soluble vitamins during blanching. This cannot be avoided.*

The vegetables which have been blanched and allowed to dry on a towel and those which did not need blanching now are ready for the actual drying operation. This step is initiated by arranging the food efficiently in a single layer, with no areas overlapping, on an appropriate screen rack. If the food is to be oven dried, it is feasible simply to place netting or cheesecloth over the existing rack and fasten it firmly before placing the food on it.

**Sun drying**

Foods to be sun dried should be positioned in their screened compartment. Careful attention needs to be given to fastening the upper screen tightly in place so that there will not be an opporunity for insects to enter the compartment during the extended period of time the food will be outside.

For sun drying, a protected area with maximum sunshine is essential. There should be no possible environmental problems, including pesticide sprays and animals or birds. If possible, it is an additional measure of protection for the food to place the drying compartment well above the ground, perhaps on a picnic table. The drying period for sun dried foods well may be three or four days. The food being dried should be turned over once daily. If there is any likelihood of dew or precipitation at night or if the temperatures at night are more than 20°F cooler than the days, the drying trays should be brought indoors at night. This prevents the addition of moisture and hastens the drying process.

**Oven drying**

For optimal oven drying, good air circulation is needed. A porous net, a relatively coarse screen, or trays at least two inches smaller than the oven may be used to help promote good circulation. A maximum load for efficient drying of food in most ovens is between four and six pounds. As many as four trays of food can be dried in an oven at the same time. Prior to actually placing the trays in the oven, the oven should be preheated. This eliminates the problem of exposing the racks of food nearest the heating areas to too high a temperature. For optimal use of electrical energy or gas, the oven door should be left closed, and the thermostat should be set at 140°F. By leaving the door closed, a minimum of heat will be lost from the oven, yet the oven venting is adequate to remove the moisture if the food is stirred frequently throughout the drying period.

An alternative approach to oven drying is to leave the oven door open throughout the drying period. This has the advantage of ready venting of the moisture-laden air from the drying food, but the disadvantage of frequent cycling on of the heat source in the oven because of quick loss of heat from the oven. This method requires more energy than is required with the door closed. Another problem with this approach is that the food

Figure 3.26. The racks used for oven drying should be at least two inches shorter, both in width and depth, than the oven to enable the hot air to circulate freely.

near the bottom of the oven will be subjected to rather intense heat periodically as the oven cycles on frequently to maintain the thermostatic setting. This may cause some of the pieces to become slightly burned unless the food is watched carefully near the end of drying.

For optimal results in oven drying, the food being dried should be checked occasionally and turned as necessary to promote uniform drying throughout the food. This is true regardless of the approach to oven drying which is being used.

**Dehydrator**

Dehydration in a dehydrator is a simple process. After the food has been placed efficiently on the trays, the trays are arranged in the dehydrator. It may be necessary to rearrange the trays occasionally if a natural-draft dehydrator is being used because drying ordinarily does not proceed uniformly in this type of dehydrator. Dehydrators equipped with a fan may not require rearrangement, but it is wise to check for uniformity of drying while determining the effectiveness of the drying operation in a new dehydrator. Even when good heat distribution exists in a dehydrator, the food should be turned over on the trays to promote more uniform drying of the pieces of food.

Precise times required for drying cannot be given since conditions are somewhat variable in drying operations. Not only are temperature and humidity variable, but also the moisture content and thickness of the food vary. Consequently, the times suggested for drying in Tables 3.1 and 3.2 are intended only as guidelines, not as absolutes. Most vegetables should be dried until they feel distinctly dry and are brittle. For example, onions, green peppers, potatoes, squash, corn, and beans will be quite brittle when dried sufficiently for good storage. In contrast, however, carrots will be distinctly leathery when they are dried enough. Red peppers will become dark in color and will be pliable

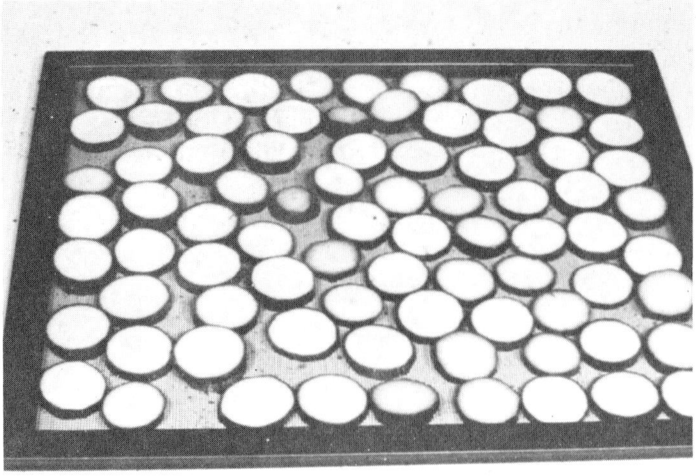

Figure 3.27. Drying racks should be arranged efficiently to permit maximum efficiency of shelf capacity. The food to be dried can just about touch other pieces, but air needs to circulate through the trays of food.

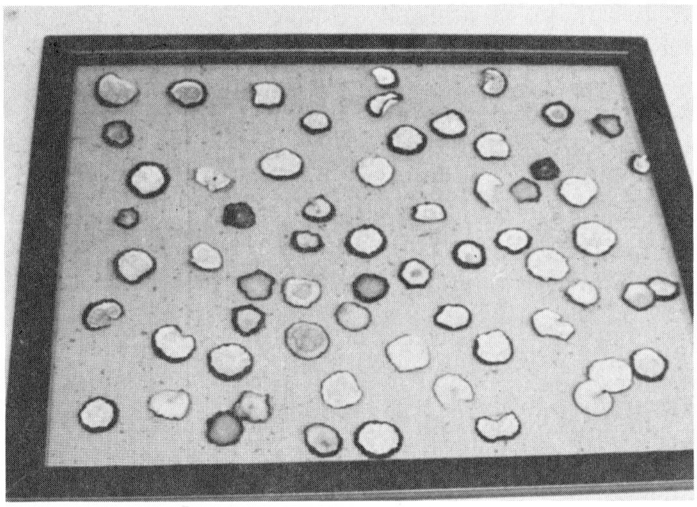

Figure 3.28. Dried foods occupy a smaller volume than did the fresh product. Note the difference in the area covered by these dried slices of zucchini and the fresh slices pictured in Figure 3.27.

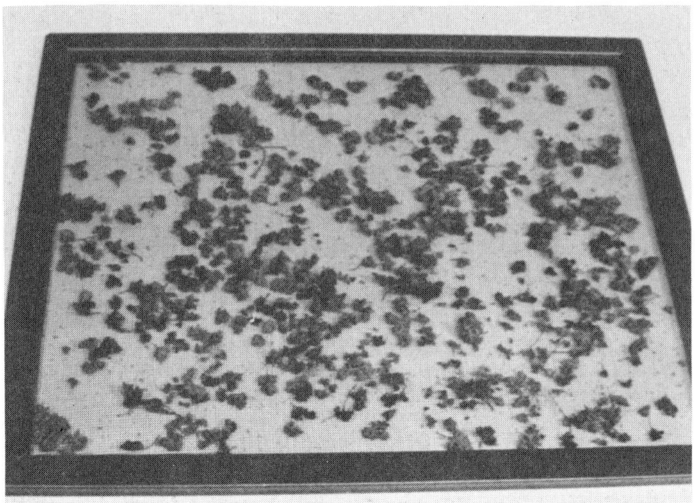

Figure 3.29. Parsley is quite brittle when it has been dried sufficiently for optimum storage.

enough to be bent without snapping. For maximum shelf life of the dried food, it is important to dry sufficiently, usually to approximately 10 percent moisture content. Testing for doneness should be done after the sample has had a chance to cool. Otherwise, the food will be somewhat more pliable than it will be when cooled, thus suggesting that the food is not as dry as it actually is.

## Fruits

### Preparation for drying

The preparation for drying fruits is influenced by the color of the fruit and by the drying technique which will be used. Blanching is sufficient preparation for fruits which will not have detrimental color changes during drying. For fruits with pigments undergoing unfavorable changes during drying, some type of treatment, either in a solution or sulfuring, is recommended to minimize the undesirable change. Sulfuring is not recommended for fruits which are being prepared for oven drying.

Some fruits, notably grapes, figs, plums, and prunes, need only to be blanched. Use of a solution or sulfuring only introduces an unnecessary step because their colors will remain pleasing during drying, and their flavors also will be good. These fruits will dry more quickly as a result of blanching because blanching helps to check or break the skin in several places. These checks facilitate moisture loss during drying. The use of a lye solution to accomplish this checking sometimes is suggested, but there are hazards possible with this severe treatment. A more convenient method to hasten drying is simply to use a sharp-tined fork to make some punctures in the skins of fruits to be dried.

Convenient techniques can be used for minimizing color change in fruit; these include soaking fruit in a sodium sulfite or sodium bisulfite solution, in an ascorbic acid solution, in an acidic fruit juice (lemon juice or canned, unsweetened pineapple juice), or in a salt solution. Sodium bisulfite or sodium sulfite can be purchased from a pharmacist. The solution is prepared, using a tablespoon of the sulfite or bisulfite salt per gallon of water, and the fruit is soaked for 10 to 15 minutes before being drained and placed on racks for drying. The ascorbic acid solution utilizes 1½ tablespoons of the vitamin per gallon of water. As with the bisulfite treatment, the soaking period is 10 to 15 minutes. A comparable soaking period is used for either the fruit juice treatment or a salt solution. The salt solution, perhaps best suited to treating apples, is prepared using between four and six tablespoons of salt per gallon of water to prevent browning. The drawbacks to these soaking treatments include loss of some water soluble vitamins and some water logging of the fruit (which slows the drying process).

The techniques outlined above are recommended for oven drying of fruits, but sulfuring is an excellent procedure for fruits being dried by other techniques. Sulfuring aids in retention of nutrients and promotes an attractive appearance of dried fruits. Preparation for sulfuring is done by pitting or cutting. Apricots are a familiar example of a fruit which is sulfured prior to drying. The cut fruit is placed with the cut surface upward on wooden trays. Metal is corroded by the fumes during sulfuring so non-metallic materials are the only suitable equipment to serve for racks. There must be adequate opportunity for the fumes from the burning sulfur to reach all of the fruit. Therefore, there needs to be sufficient rack space for the pieces of fruit to be arranged in an uncrowded placement in single layers on a tray. However, several trays can

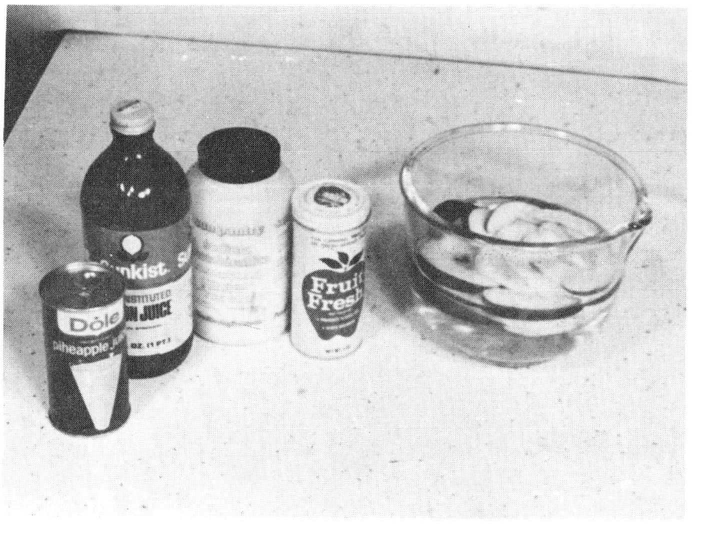


*Figure 3.30. Color changes in some fruits during preparation and drying can be avoided or retarded by using one of these techniques: (left to right) pineapple, lemon, or other acidic fruit juice; sodium bisulfite (dissolved in water); or ascorbic acid (dissolved in water) Apples sometimes are soaked in a salt solution as yet another means of preventing browning.*

*Figure 3.31. By slicing bananas directly into a strainer immersed in pineapple juice, the fruit immediately is protected from browning by the acidic medium. The slices can be lifted easily from the juice by lifting the strainer when all of the fruit is prepared.*

*Figure 3.32. Cleaning up problems can be reduced by placing a jelly roll pan under the rack while the moist slices of the acid-dipped bananas are placed on the rack for drying. When no more dripping occurs, the rack can be transferred to the oven or dehydrator.*

be stacked, using blocks or bricks, to make it possible to sulfur up to 40 pounds of fruit at one time.

The sulfur used in sulfuring is sulfur flowers (U.S.P. standard). This is available at pharmacies. There also are sulfur preparations for garden dusting, which can be used. Garden preparations must contain at least 95 percent sulfur, and even at that level of sulfur, ignition is difficult. The flowers of sulfur are easier to use. The amount of sulfur required for sulfuring fruits is one tablespoon of the sulfur for each pound of fruit. This material is placed in a metal can or disposable aluminum foil container positioned at the edge below the bottom rack.

The final piece of equipment for sulfuring is a large cardboard box, which is large enough to fit over the stacked racks with at least an inch to spare on all sides and above the fruit. A slash needs to be made for about four inches near the bottom of the box, on one side only. A second slash is made in the top of the box near the far edge away from the original slash. These slashes can be used to help circulate the fumes through the box.

When everything is in readiness, the sulfur is ignited with a match, the box is placed in position, and timing of the sulfuring begins. The time for sulfuring is quite variable, as can be seen from Table 3.2. Times recommended vary from 45 minutes for apples to five hours for pears. The fruit is left in the sulfuring chamber, without disturbance, until the appropriate length of time has elapsed. Then the fruit is removed and placed immediately on drying racks for the drying process. When fruits have been sulfured enough, they will have a little juice gathering in the cavity left by the pit, and the surfaces will be lustrous.

**Drying**

As soon as the special procedures for controlling browning of fruits have been completed, the fruit should be placed on racks, and drying is started. For efficient drying, fruit should be placed in a single layer on the rack without the pieces of fruit touching. Drying can be done by sun drying, oven drying, or in a dehydrator. The precautions and suggestions presented above for vegetables apply to drying of fruits. However, fruits will be leathery and pliable when they are dried, rather than being brittle like most of the vegetables. Time for drying is influenced by several factors, including the moisture content of the original fruit, the mode of drying, the environmental conditions, and the size of pieces used.

## Specialty Products

Usually home drying includes vegetables and/or fruits as the major items to be processed. However, dried meat or jerky is increasing in its popularity. Only very lean meat is used because the fat will shorten the shelf life by becoming rancid. Meats to be dried are cut into very thin strips, a task which is accomplished most easily by partially

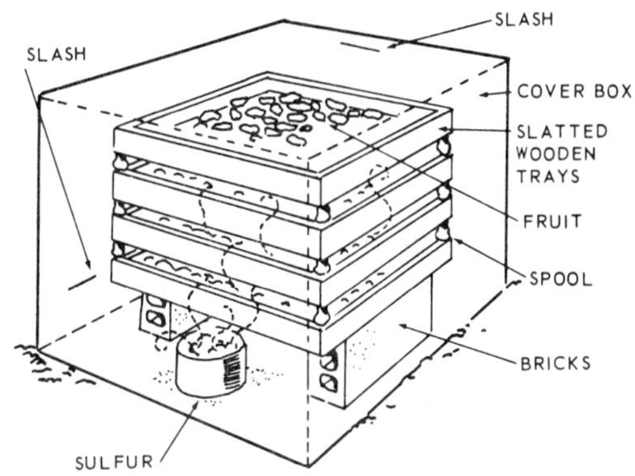


*Figure 3.33. Diagram of the basic design of equipment used in sulfuring.*

*Figure 3.34. Dried apples are leathery when they are dehydrated sufficiently.*

freezing before slicing so that the meat is semi-rigid. If desired, flavors can be added by marinating prior to drying. Beef, turkey, chicken, and salmon can be dried into very acceptable jerkies. Their drying is facilitated by the use of a rather salty marinade. Jerky can be made very satisfactorily in the oven, but sun drying or even drying over very low coals outdoors can be done. Care must be taken to avoid contamination outdoors. Jerky can be used in casseroles and soups, but it often is preferred simply as a finger food.

Fruit leathers represent a special type of dried fruit product, for this is a puree of fruit, rather than individual pieces. The fruits can be pureed easily in a blender, with lemon juice or ascorbic acid (one tablespoon or one-fourth teaspoon, respectively) being added to each quart of fruit. A small amount of honey or sugar can be added, if desired. The pureed mixture then is spread to a depth of one-fourth inch in a jelly roll pan lined with plastic wrap. Drying can be done by any of the three methods. When the product has dried sufficiently, the surface will feel firm to the finger and will be able to be pulled without tearing from the plastic. Storage is accomplished by simply rolling the leather and wrapping the roll in plastic wrap or aluminum foil.

## STORAGE

One of the advantages of dried foods is the convenience in storage. Foods to be stored are cooled to room temperature after being removed from the dryer before being packaged in plastic bags or other containers which can be sealed tightly. Tight closure is essential so that the food will remain clean and will not take up moisture from the air during shelf storage. If long-term storage is anticipated, sealed plastic bags can be placed in larger containers or boxes to keep the bags from becoming dusty. The use of fairly small plastic bags for storing dried foods is recommended. By storing small quantities of food, only a limited amount of food might be lost if mold or other form of spoilage should happen to develop in one of the bags.

Ideally, dried foods should be stored in a cool, dark, dry cupboard for later use. Refrigeration and/or freezing are not necessary, although freezer storage can be used to extend shelf life, if desired. Sun dried foods, however, should be stored in the freezer for at least two days immediately following the drying period to provide insurance against possible insect contamination.

Shelf life of dried foods varies with the food and the storage conditions. A rule of thumb is that most dried foods can be stored for a year at room temperature, providing the storage area is cool. However, fruit leathers maintain optimum quality with room temperature storage for a period of only about a month. Freezer storage extends this to a year. Jerky also has limited shelf life because the fat remaining in the meat will become rancid. As a general guideline, storage is improved if the dried foods are placed in plastic bags before packaging in any other containers.

Careful labeling of packages of dried foods is important for inventory purposes so that dried foods can be consumed within the one year time period. Labeling should include the date of drying and any other information pertinent to the use of a particular food.

## SERVING DRIED FOODS

Dried foods are such convenient finger foods that many of them are likely to disappear as portable snack items. However, dried foods can be used in several ways in meals, too. Dried vegetables are best used in stews, casseroles, soups, and other dishes where they will not be featured alone. The

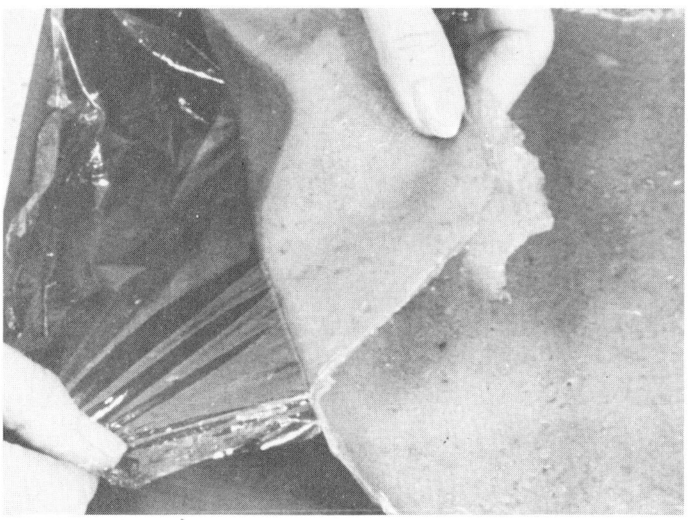

*Figure 3.35. Fruit leather derives its name from the leather-like quality of the dried product. Sometimes leathers are referred to as fruit rolls or fruit taffies.*

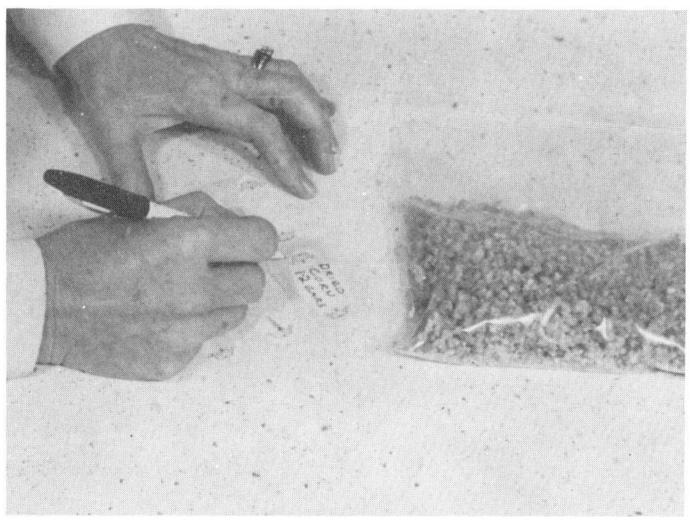

*Figure 3.36. Dried foods should be labeled carefully with the name of the item and the date prior to being stored.*

flavor of the rehydrated vegetable can be heightened by adding a bit of seasoning or some onion. Vegetables can be soaked to rehydrate them, but it often is more practical simply to place them in the food in which they will be cooked. It is helpful, however, to soak root and stem vegetables for at least half an hour before adding to other foods. Dried legumes require less cooking time if they are soaked several hours before cooking.

Dried fruits often are used in desserts or simply as the stewed fruit. Rehydration by simmering dried fruits in water until they are plump and tender is the appropriate procedure both for preparing stewed fruit and for fruit to be used in cookies and other baked products. If dried fruits are used in batter and dough products without being rehydrated, they will absorb moisture from the dough and modify the characteristics of the product in which they are incorporated.

### NUTRITIVE VALUE

Vitamin retention in dried foods is related to the moisture level and the storage temperature. To illustrate (Kramer, 1974) reports reveal a 10 percent loss of ascorbic acid with moisture at a level of one percent, but a 30 percent loss in ascorbic acid was found under the same storage conditions when tomato flakes had a moisture level of five percent. When dried onions were stored at 70°F., they lost 10 percent of their vitamin C in 1 month, but three months of storage were required for 10 percent loss of this vitamin when storage temperature was reduced to 55°F.; at 48°F. 10 percent loss of vitamin C did not occur until the onions had been stored for six months. Tight storage containers and storage at are cool temperature obviously are important to the retention of this labile vitamin. Refrigerator storage greatly enhances the retention of vitamin C. As much as half of the content of vitamins A and C is reported to be lost in six months storage of dehydrated vegetables stored at 68°F., but only about 10 percent loss occurs when storage is at 32°F. (Daoud and Luh, 1967).

Protein content also is of interest because availability of specific amino acids varies as a result of dehydration and storage. Methionine availability remains high, for example, but lysine availability is reduced, particularly under poor conditions of storage.

## SUMMARY

Fruits, vegetables, and sometimes meats are dried. Preparation for drying includes careful washing, removal of any spoiled areas, and special treatment (either sulfuring for sun drying of most fruits or blanching for oven drying) for fruits. Methods of drying include sun drying, oven drying, and in a dehydrator. Foods are dried until they are either brittle or leathery, a texture found in foods containing very low moisture levels. Storage is done in air tight containers, usually plastic bags, in a dry, cool cupboard. Packages should be labeled with date of drying so that they will be used before they have been stored more than a year. Dried foods may be used in their dried form or rehydrated by soaking or simmering in water.

## BIBLIOGRAPHY

Burr, H.K. and R.M. Reeve. Potatoes. In *Food Dehydration. Vol. 2.* ed. by W.B. Van Arsdel, et al. Avi Publishing Co. Westport, Conn. 1973.

Daoud, H.N. and B.S. Luh. Packaging of food in laminate aluminum-film combination pouches. *Food Tech. 21 (3A):* 21. 1967.

Feinberg, B. Vegetables. In *Food Dehydration, Vol. 2.* ed. by W.B. Van Arsdel, et al. Avi Publishing Co. Westport, Conn. P. 1. 1973.

Kennedy, S.O. *How to Dry Food for the Home.* Bulletin C166. Cooperative Extension. University of Nevada, Reno, Reno, Nevada. 1975.

Kramer, A. Storage retention of nutrients. *Food Tech. 28:* 50. 1974.

Miller, M.W., et al. *Drying Foods at Home.* Leaflet 2785. Cooperative Extension. University of California. Berkeley, California. 1975.

Miller, M.W., et al. *Drying Foods at Home.* Home and Garden Bulletin No. 217. U.S. Dept. Agriculture. Washington, D.C. 1977.

Sulzbacher, W.L. Meat dehydration. In *Food Dehydration. Vol. 2.* ed. by W.R. Van Arsdel, et al. Avi Publishing Co. Westport, Conn. P. 246. 1973.

Van Arsdel, W.B., et al. *Food Dehydration, Vol. 2.* Avi Publishing Co. Westport, Conn. 1973.

conditioning—put in bag + moisture well distribute

# Chapter 4. Recipes for Drying

## VEGETABLES

All vegetables to be dried should be washed gently, yet thoroughly in cold water in preparation for drying. This step is essential for removing any dirt, foreign particles, and sprays which may be on the food. Then the vegetables are prepared by removing any blemished areas and woody portions. Most vegetables also are sliced before being dried. Specific preparations for several vegetables which do not require blanching are outlined below. Other vegetables require blanching before drying.

Drying is done in the sun (with the food protected from insects), in the oven (at 140°F), or in a dehydrator. To help speed drying, the food should be turned occasionally. Testing for completion of drying can only be done with reasonable accuracy when a sample has been allowed to cool to room temperature before being tested.

Storage of dried foods should be done as soon as the foods have cooled. Plastic bags, tightly fastened, provide convenient storage containers. These bags can then be placed in a larger, more rigid container for long-term storage in the cupboard. Glass jars are excellent containers for dried foods if they are stored in a dark place. All dried foods should be identified with the name of the product and the date of drying.

### Drying without Blanching

#### Chives

Chives are washed, patted dry, and then snipped with shears into short segments (about ¼-inch long). Drying is continued until the pieces are brittle.

#### Herbs

Fresh herbs are washed well, patted dry with paper toweling, and then dried. When they have been dried sufficiently, the leaves will be brittle and can be removed easily from the stems. Crumble the leaves or let them remain in coarse pieces, depending on personal preference. Storage is done for 24 hours in uncapped glass jars. If moisture collects on the jars, the herbs need additional drying time. After thorough drying, storage is done in closed containers, as for other dried products.

#### Horseradish

The vegetable is scrubbed well, peeled, and grated for drying. Drying is done until the pieces are brittle.

#### Mushrooms

Care should be taken to be certain that only edible mushrooms are dried. Drying does not make

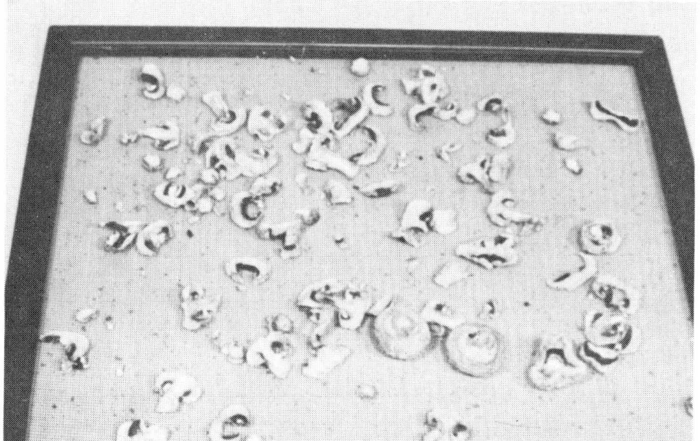

*Figure 4.1. Mushrooms may be left whole or sliced for drying. Drying is completed when mushrooms are dry and leathery.*

poisonous varieties safe to eat. Mushrooms need to be washed thoroughly in preparation for drying to be sure that all of the clinging dirt has been removed. Then any woody areas are cut away, the stem is trimmed, and the mushrooms are sliced, if large. Drying is continued until the slices or pieces are very dry and leathery.

**Okra**

After a thorough scrubbing, okra is sliced crosswise in slices about ¼-inch thick. This vegetable is dried until leathery.

**Onions**

Onions are prepared for drying by washing, peeling the outer, parchment-like covering, and cutting off the stem and root ends. For drying, onions may be chopped or sliced in ¼-inch thick slices. They are dried, turning at least twice, until they are brittle.

**Parsley**

Parsley is separated into clusters after being washed and drained. The tough portions of stems are removed. Drying is done until the parsley leaves are brittle and flaky.

**Peppers and pimientos**

The stems, core, and all of the seeds need to be removed from peppers and pimientos after they have been washed. These vegetables can be cut into small squares or sliced in ¼-inch thick slices prior to drying. Drying is completed when the pieces are brittle.

**Tomatoes**

Tomatoes may be peeled easily for drying by dipping them briefly in boiling water to loosen the skins. However, it is also appropriate to prepare tomatoes for drying without peeling them. The core is cut out, and the tomatoes are cut into slices half an inch thick or in wedges, if preferred. The dried tomatoes should be dry and slightly leathery.

## Drying following Blanching

**Techniques for blanching**

Blanching may be done either in steam or in water. This process needs to be done long enough to heat the vegetable throughout so that enzymes which could cause deteriorative changes will be inactivated.

Devices for steam blanching require some type of rack to hold the vegetables in active steam above boiling water. Suitable choices include steamers, colanders which will remain suspended above the boiling water, a strainer, a piece of cheesecloth encompassing a domed lid, and an aluminum foil pie plate of an appropriate size to fit above boil-

*Figure 4.2. Onions provide another example of a vegetable which is dried without blanching. They may be sliced or chopped. The onions on this tray were chopped coarsely prior to drying.*

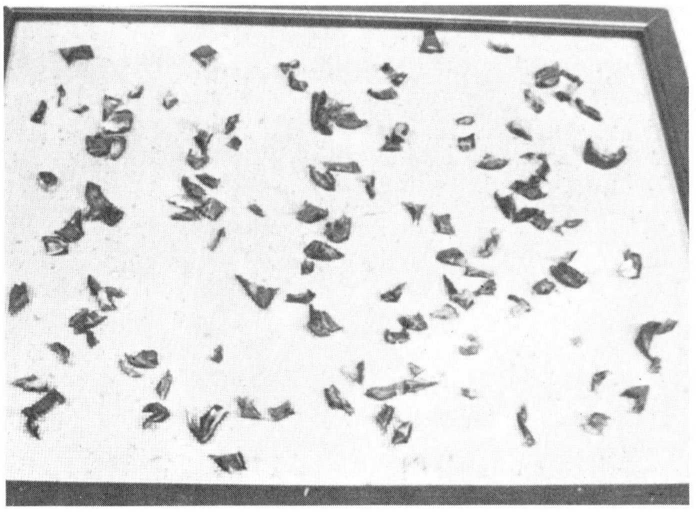

*Figure 4.3. Peppers are seeded and then cut into moderate to small pieces or sliced thinly for drying. These pieces of pepper have been dried until they are brittle.*

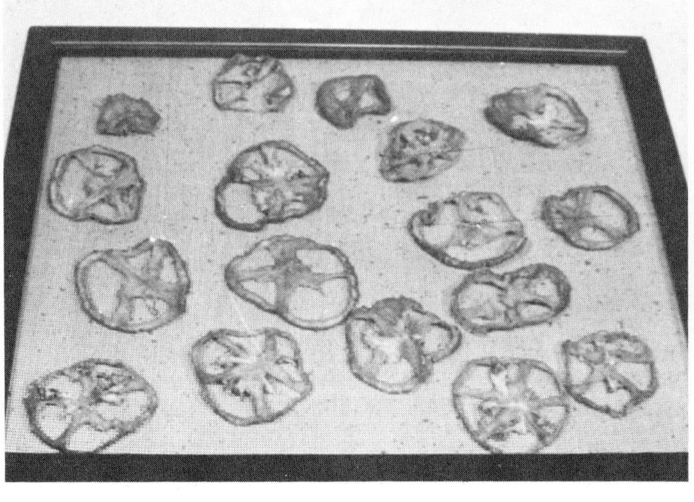

*Figure 4.4. Dried tomato slices have a somewhat delicate interior as a result of the drying process. For best results, varieties of tomatoes with a meaty interior (Beefsteak, for example) are recommended for drying.*

ing water in a saucepan and in which perforations have been made to allow free flow of the steam.

Relatively tight-fitting lids are important for steam blanching, since it is important to have the steam trapped around the food as much as possible. Steam blanching requires almost twice as long a blanching period as is required when blanching is done in boiling water.

For blanching in water, it is necessary to have a wire basket, a net bag, or a strainer so that the vegetables can be dipped into the boiling water and removed quickly. Blanching is done when the vegetable is hot throughout and the inner cells are somewhat translucent, yet not tender.

*Figure 4.5. The tight-fitting lid and large surface area of a roaster make this a useful piece of equipment for steam blanching.*

*Figure 4.6. The kettle of a water bath canner can be used conveniently for blanching large quantities of vegetables in boiling water.*

**Asparagus**

Wash thoroughly and cut into pieces about an inch long. Large pieces should be cut in half lengthwise. Dry until somewhat leathery.

**Broccoli**

Wash broccoli very well, cut into short pieces. Blanch. Dry until flowers are brittle and stalk is leathery.

**Brussels sprouts**

Wash. Cut in half lengthwise, removing any stem and the outer leaves, if wilted or blemished. Blanch. Dry until somewhat leathery.

**Cabbage**

Wash. Remove the core. Cut into strips approximately ⅛-inch wide. Blanch. Dry until somewhat leathery.

**Carrots**

Wash, pare, trim root and top, and cut into slices or strips about ⅛-inch thick. Blanch. Dry until leathery.

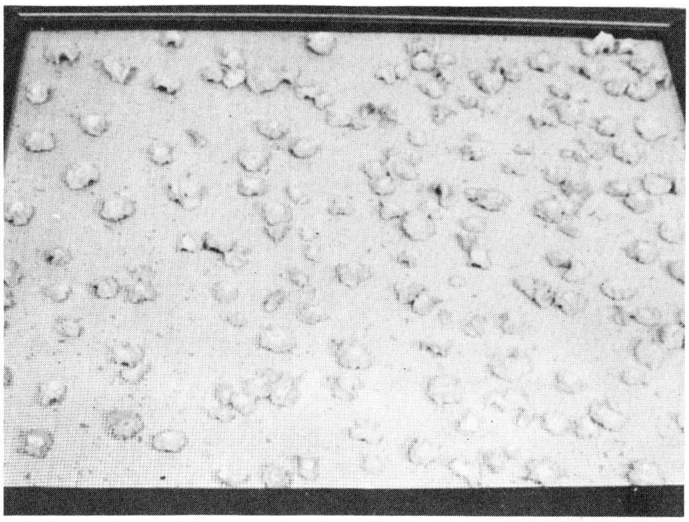

*Figure 4.7. When slices of carrot dry, the edges split in several places, making an attractive pattern. The vegetable is an example of one requiring blanching for optimal performance during dehydrated storage. The texture of carrots becomes leathery.*

**Cauliflower**

Wash well and break into flowerets. If flowerets are thick, slice in half. Blanch in boiling water (steam blanching results in a stronger-flavored product). Dry thoroughly.

**Corn**

Husk the ears, wash well, and blanch. Use a sharp knife to cut the kernels from the cob. Dry until hard.

**Celery**

Wash thoroughly to remove all dirt. Trim the stalks and leaves. Slice crosswise or diagonally into slices about ¼-inch thick. Blanch prior to drying. Dry the slices until somewhat leathery. The leaves are dried (if desired) until brittle.

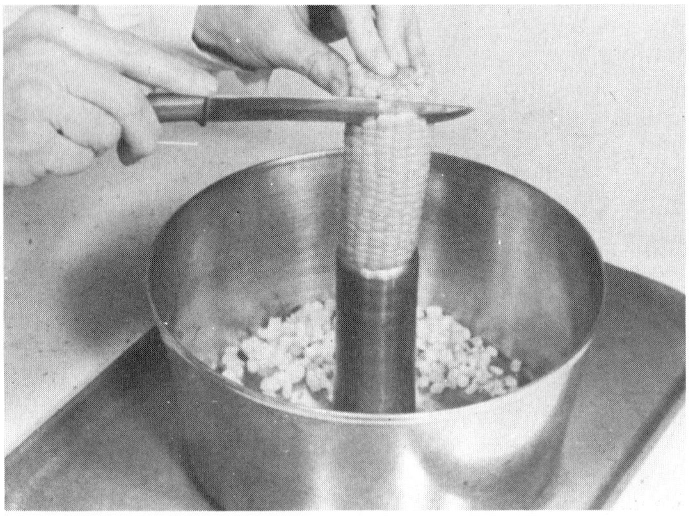

*Figure 4.8. Following blanching, sweet corn is cut from the cob in preparation for drying. An angel cake pan is convenient as a support for the cob and a collection device for the kernels as they fall from the cob.*

**Greens (including spinach)**

Wash very carefully. Remove coarse stems and any bruised leaves. Blanch until wilted thoroughly. Dry. Leaves will be brittle when drying is finished.

**Peas**

Shell the peas and wash them prior to blanching. Dry well until dry in center and hard.

**Potatoes**

Wash potatoes well, peel carefully, and slice into ¼-inch thick shoestrings or ⅛-inch thick slices. Blanch before drying. Dry until leathery.

**Summer squash and eggplant**

Wash and peel. Cut into crosswise slices about ¼-inch thick. Blanch. Dry until leathery.

**Winter squash**

Wash and peel squash. Remove seeds and coarse strings in the center cavity. Slice in strips between ⅛ and ¼ inch in thickness. Blanch and dry. Dry until leathery.

## FRUITS

Fruits to be dried should be fully ripened. They need to be washed very thoroughly to remove dirt and sprays which may be on the fruit. Any areas which are bruised or spoiled must be removed prior to drying. Fruit may be sliced or cut into halves. Perforations or cracks in the skin will facilitate drying because they provide additional avenues for evaporation. This is particularly important when drying large pieces. Perforations can be made by puncturing with a fork, by blanching fruits long enough to rupture the skin, or by checking the skin with a lye solution (described below).

## Pretreatment

**Acid treatment**

A 1-minute dip in an acidic fruit juice, in ascorbic acid, or in a sodium bisulfite solution (one tablespoon sodium sulfite or sodium bisulfite per gallon of water) for a 10 to 15 minute soaking period will halt color changes. The ascorbic acid solution is of similar strength, i.e., one to one and a half tablespoons of ascorbic acid per gallon of water. Fruit juices, such as pineapple juice, also are effective in minimizing browning of fruits to be dried.

**Sulfuring**

Pretreatment of fruits is an important aid in retaining color and flavor in dried fruits. Sulfuring often is the method to be preferred for quality, but it is less convenient than using ascorbic acid or fruit juices to prevent oxidation. For oven drying, however, the sulfuring technique is not recommended because of the strong sulfur smell which will be liberated in the kitchen during drying.

**Checking with lye**

Another pretreatment is to check the skin, thus promoting more rapid drying. Puncturing the skin with a fork is a method often preferred because of its greater ease. However, lye treatment is very effective and sometimes is done in preparation for drying prunes.

The lye solution is made by stirring a tablespoon of lye into two quarts of cold water (using a wooden spoon or stick). Caution is needed when using lye. The solution is brought to a boil before the fruit is dipped very briefly into the solution. Since this lye solution is dangerous to handle, care should be taken to avoid splattering. It is necessary to dispose of the solution carefully, too. This can be done by pouring the solution carefully down the sink drain and rinsing thoroughly. Rubber gloves provide useful protection for the hands when handling lye solutions. Of course, the fruit dipped in lye solution needs to be rinsed very thoroughly under running water before drying. The lye treatment will leave many small breaks in the skin of the prunes, thus promoting more rapid evaporation and drying.

Because of the hazards presented by the lye solution, the following suggestions should be noted and observed carefully:

1. Use only stainless steel or unchipped enamelware for making and heating the lye solution
2. Wear rubber gloves when working with the lye solution.

3. Stir only with wooden utensils.

4. Rinse the lye-treated fruit very thoroughly under running water.

5. Pour the lye solution very carefully into the sink drain and rinse the drain well.

## Additional Treatment

Since fruits may tend to stick on the drying trays, it is helpful to spray the trays with a non-sticking vegetable spray before placing the pretreated fruit on the trays. Fruits should be turned at least once or twice to facilitate drying. Drying is done until fruits are leathery.

**Apples**

Apples are washed well and cored for drying. Paring is optional. Then they are sliced into rings or slices ⅛-inch thick and immediately placed in fruit juice or an ascorbic acid solution to prevent discoloration. It is not necessary to blanch apples, but they may be blanched for five minutes, if desired.

**Apricots**

Apricots to be dried are washed, pitted, and cut in halves prior to pretreatment. They may then be punctured with a fork or sulfured. Sulfuring is recommended if color retention is important. If sulfuring is not being done, a water blanch may be done while the fruit still is whole. Drying follows pretreatment.

**Bananas**

Peel the bananas (one at a time) and slice into slices ¼ inch thick. Immediately dip the slices in pineapple or other acidic juice to prevent discoloration. Dry until pliable.

**Figs**

When possible, figs should be allowed to remain on the tree until they drop. They may be dried whole without any pretreatment before drying. When dry, they will be flexible and a little sticky.

**Grapes**

Seedless grapes are preferred for drying. Preparation requires only thorough washing and removal from the stems. No pretreatment is required, but puncturing the skins with a fork will speed the drying process.

**Peaches and nectarines**

Blanch or sulfur the fruit. For blanching, leave the fruit whole. Peel after blanching (if desired), halve, and pit. For sulfuring, halve and pit fruit before sulfuring. Sulfuring is particularly helpful in maintaining good color in the dried fruit.

**Pears**

Following washing, pears are cut in half, cored, and pared, if desired. The fruit needs pretreatment by sulfuring or steam blanching. Ascorbic acid or fruit juice can be used to minimize color change. The fruit is dried until pliable, but no longer moist in the center.

**Prunes**

Prunes may be prepared for oven drying simply by washing the fruit well in tap water and patting dry before placing the whole fruit in the oven. Sulfuring is recommended as a means of enhancing the flavor of the dried product. Drying is facilitated by puncturing the skins with a fork or by checking the skins with lye, as described above.

## Fruit Leather

Fruit leathers, also called fruit taffies or fruit rolls, are basically pureed fruits which have been sweetened and dried. Almost any fruit or combination of fruits will give flavorful and pleasing results. Fruits suggested for use in leathers include: apricots, apples, grapes, cherries, berries of all types, pineapple, bananas, oranges, peaches, pears, plums, and papayas. Apples often are used with other fruits because they help to contribute the desired leather-like quality, a result of the high pectin content of apples.

Recipes for fruit leathers can be varied to present a wide range of flavors. The fruits can be enhanced by adding flavorful ingredients, such as cinnamon, nutmeg, grated lemon or orange peels, raisins, coconut, and nuts. The flavor of the mixture prior to drying is similar to the flavor in the finished product. Consequently, the mixture can be modified to taste prior to drying.

**Drying techniques**

Fruit leathers may be dried in the oven or in the sun. Oven drying is easier to control because it requires less drying time and presents less likelihood of insect contamination. Oven drying of the leather is done simply by putting the jelly roll pan of puree in the oven, setting the oven at 140 to 150°F, and letting it dry at that low setting with the oven door closed. When the upper surface is no longer sticky, the leather can be peeled from the plastic wrap, and the sheet of leather is flipped over to expose the second surface to complete the drying process efficiently. As soon as both surfaces are no longer sticky, the leather is done. If the leather is dried too much, the product will be too brittle to roll. This situation can be remedied by dampening paper toweling very slightly, placing it on the leather, and letting it stand just long enough for the leather to absorb enough moisture from the paper to restore the pliable nature of the leather. A comparable process is used to dry leathers in a dehydrator.

Sun drying requires a level surface in the full sun. Cheesecloth needs to be stretched above the drying leather to protect the fruit from insects. However, it is important to keep the cloth from touching the surface of the leather. The length of time required for sun drying will vary considerably, but often more than a day will be needed. The drying leather should be brought indoors at night and returned to the sunlight in the morning. When it is possible to handle the leather, remove the leather from the plastic wrap and turn the leather over to dry the second side. If the leather becomes too dry, soften as described above before wrapping the rolled leather for storage.

*Figure 4.9. Fruit to be made into leather is washed and drained well before being placed in the blender. Any imperfections are removed at this time. Strawberries are prepared by removing their hulls. The other ingredient needed is honey.*

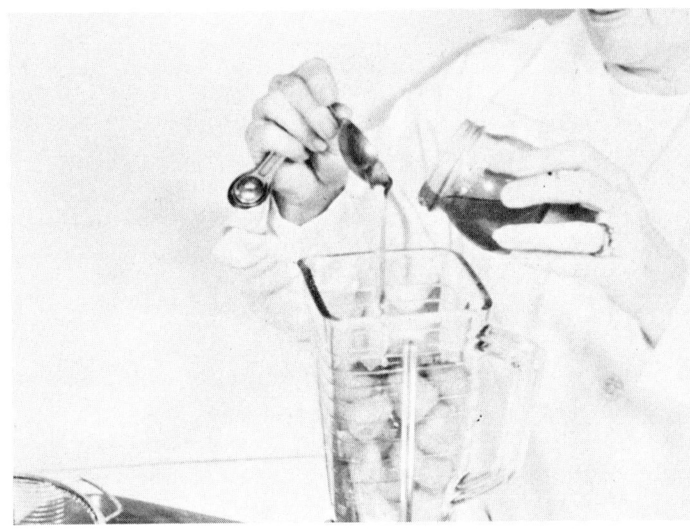

*Figure 4.10. The addition of honey to the pieces of fruit in the blender aids the blender in pureeing the fruit easily.*

**Recipe for fruit leather**

1. Wash fruit well and drain thoroughly.
2. Cut into large pieces and add to blender gradually. Add enough fruit to make two cups of pureed fruit. Note that the use of the blender makes it unnecessary to pare the fruit, but seeds must be removed.
3. Add approximately one to two tablespoons of sugar or honey to sweeten to taste.
4. To prevent discoloration of light-colored fruits, stir in two teaspoons of lemon juice.
5. Line a jelly roll pan (10 x 15 x 1") with a piece of plastic wrap which is taped to all sides of the pan.
6. Pour the pureed fruit mixture into the lined pan and spread uniformly into a layer ¼ inch deep, being sure all areas are filled.
7. Dry until the surface is no longer sticky. Then remove plastic wrap from under the leather.
8. Cool on the kitchen counter (replacing it in the pan after removing the plastic wrap).
9. Roll the cooled leather into a cylindrical shape, wrap in plastic wrap, and store in a cool, dark cupboard.

*Figure 4.11. Fruit is added gradually and pureed until a total of 2 cups of puree has been prepared.*

**Uses**

Fruit leathers can be used without modification for snack items. They also can be snipped or chopped into pieces for addition to cookies, cakes, breads, pie fillings, toppings, sauces, and salads. Another use is in the preparation of fruit beverages, which can be made by whirring five parts of water to one part of fruit leather in an electric blender.

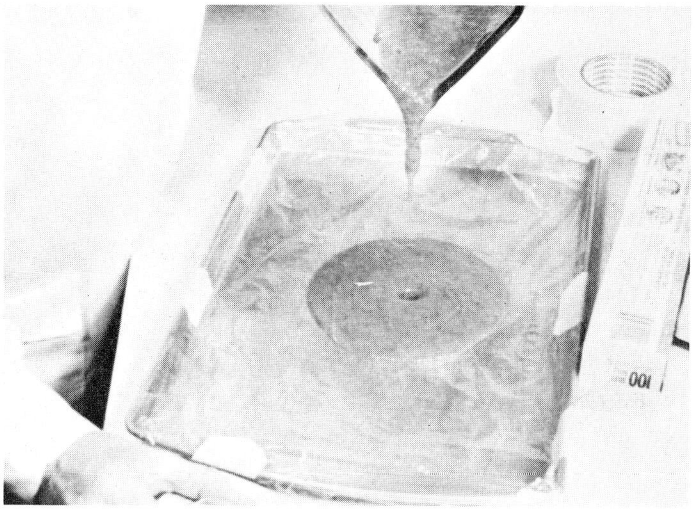

*Figure 4.12. A jelly roll pan (10 x 15 x 1 inch) is lined with plastic wrap, which is taped securely to all sides of the pan. Then the puree is poured in.*

*Figure 4.13. The easiest way to obtain uniform distribution of the puree is to tilt the pan, being sure to have the mixture run into all corners.*

*Figure 4.14. The leather now is ready for drying in the oven or dehydrator. The pan is placed directly on the rack.*

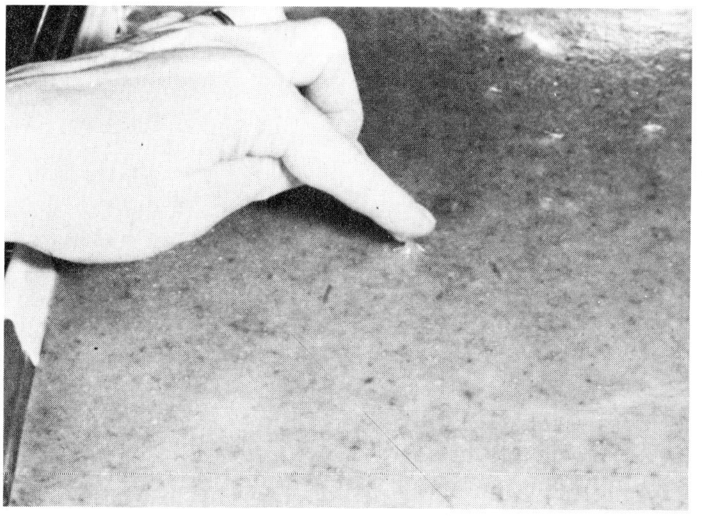

*Figure 4.15. The leather is dried until it can be touched lightly with the finger without the leather sticking. If the leather sticks to the finger, as in this illustration, more drying time is needed.*

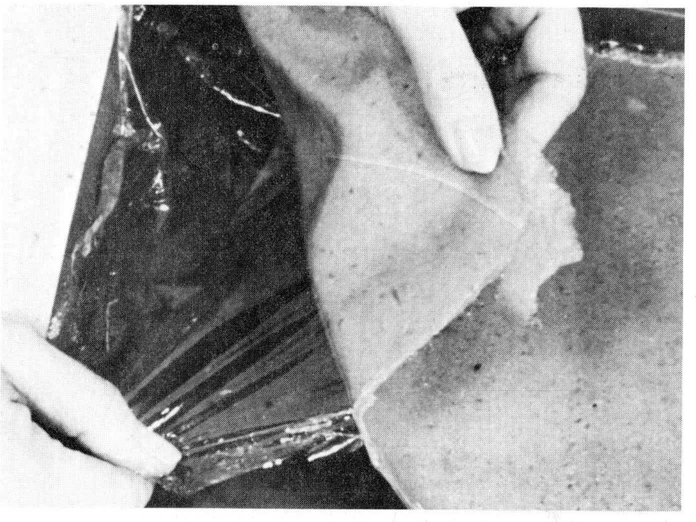

*Figure 4.16. As soon as the leather is dried, the pan is removed from the drying area, and the plastic wrap is peeled from the leather. The leather then is allowed to cool in the pan on the counter.*

*Figure 4.17. When cooled, the leather is rolled into a log, wrapped in plastic wrap, and stored for future use.*

### Vegetable Leathers

Although done less commonly, vegetable leathers also can be made. These are particularly well suited to mixtures suitable for soup bases, for flavorings in casseroles, or other hearty recipes. Vegetables are blanched and then pureed in preparation for vegetable leathers. If time to prepare for drying is not available conveniently, vegetables (and also fruits) may be pureed in amounts of two cups and frozen until it is convenient to dry the puree into a leather.

### Pride of Iowa Dried Corn

Dried corn assumes a special quality when it has been prepared in the following way prior to drying.

1. Husk two dozen medium ears of corn (makes about 2 quarts of corn kernels prior to drying), being careful to remove the silks.
2. Wash the ears of husked corn thoroughly.
3. Cut the kernels of corn from the cobs, using a sharp knife.
4. Combine 2 quarts of corn kernels, ½ cup cream, 1 tablespoon salt, and ¼ cup sugar in a heavy saucepan and heat to a boil.
5. Lower the heat and simmer for 20 minutes while stirring frequently. Cover the pan when not stirring.
6. Drain the liquid from the corn. The liquid may be saved for use in soups and casseroles.
7. Spread the kernels of corn on drying racks or cookie sheets in a shallow layer.
8. Dry in the oven at 140 to 150°F. or in the sun.
9. Stir corn occasionally during drying to be sure that all kernels become thoroughly dry before placing in a tightly sealed plastic bag or other airtight container for storage.

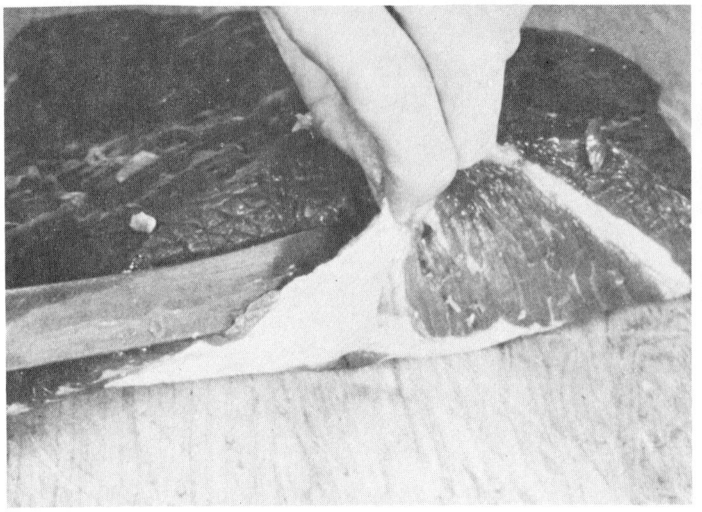

*Figure 4.18. The shelf life of dried meats (jerky) is extended by removing as much fat as possible prior to drying. This reduces the problem of rancidity which develops in fat during storage.*

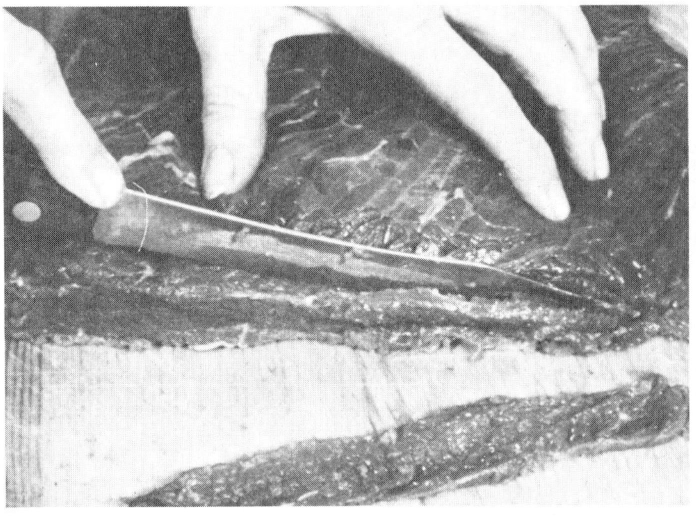

*Figure 4.19. If meat has been chilled to the point where it is beginning to freeze, it will be easier to cut the extremely thin slices needed for drying meat into jerky.*

## DRYING MEATS

### Beef or Turkey Jerky

1½ to 2 pounds lean meat (flank steak, brisket, round steak, or breast of turkey or chicken)
¼ cup soy sauce
1 tablespoon Worcestershire sauce
1 teaspoon garlic salt
1 teaspoon seasoned pepper

1. Remove all visible fat from around the muscles of the meat.
2. Cut the meat into slices between ⅛ and ¼ inch thick. This is done more easily if the meat is frozen partially so that it is somewhat rigid.
3. Blend the sauces and seasonings in a bowl to make a marinade for the meat.
4. Stir the slices of meat in the marinade to moisten all portions of the meat with the marinade. Let stand in the marinade for at least an hour.
5. With a piece of aluminum foil on the bottom oven rack to catch the drip, arrange the strips

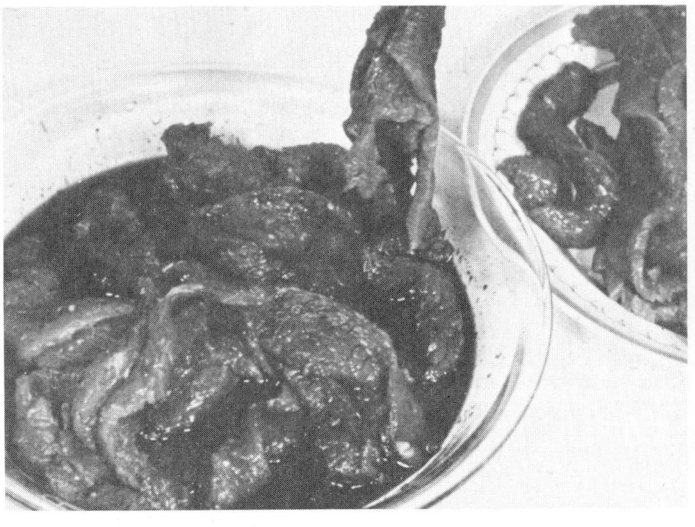

*Figure 4.20. Marinating the slices of meat for at least an hour in a flavorful marinade adds flavor interest to the finished jerky.*

*Figure 4.21. After a piece of aluminum foil has been placed on the bottom rack of the oven, the marinated strips of meat are hung separately on the bars of the upper rack for drying. The foil prevents juices and marinade from falling to the floor of the oven.*

of meat on the upper oven rack. Slide the oven racks into the oven and dry the meat at 140 to 150°F. until the meat is hard and dry (about 5 to 12 hours, depending on the thickness of the meat). Meat also can be dried in the sun, using a barbecue grill. This will require up to 4 days, with indoor storage at night.

6. After jerky is dry, use paper toweling to pat off any beads of oil on the surface. Cool before storing in plastic bags or other airtight containers. The above recipe makes about ½ pound of jerky.
7. Serve in dried form as a snack or crumble it into salads and casseroles for flavoring.

## Hamburger Jerky

1. Mix ½ pound ground beef (as lean as possible) with desired seasonings. Seasonings such as garlic powder, onion salt, soy sauce, pineapple juice, taco sauce, ground ginger, horseradish, or teriyaki sauce may be used to add flavor.
2. Perforate a piece of aluminum foil or plastic wrap with many small openings and arrange over a double layer of paper toweling.
3. Spread the meat over the foil.
4. Place a piece of wax paper over the meat, and roll the meat to a uniform thickness of ⅛ to ¼ inch, using a rolling pin. If areas do not have meat in them, fill in with some of the meat from the edges. The finished product should be a thin rectangle of meat approximately 9 by 12 inches.
5. Remove the layer of wax paper. Slide the meat with its base of paper towel and foil into the oven to dry. Dry at 140 to 150°F. until dry and brittle. Change the paper toweling when it becomes damp with drippings, and turn the jerky over occasionally when appropriate during drying.
6. Store the cooled, dried jerky in tight containers in the refrigerator or freezer. Storage for a few days can be done satisfactorily at room temperature.

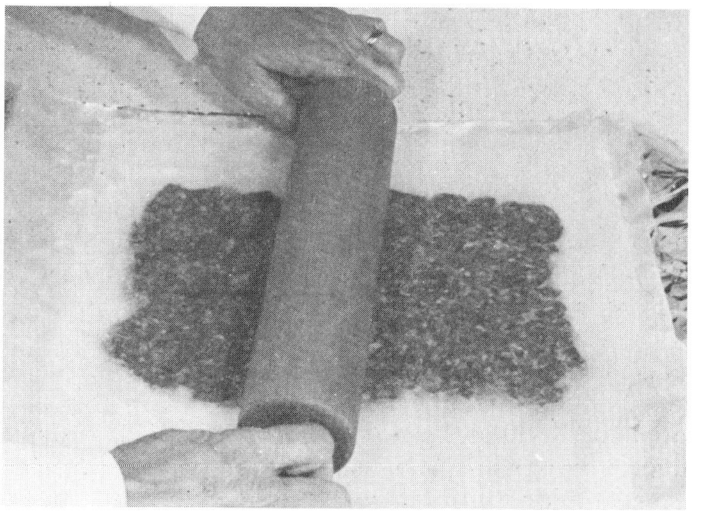

*Figure 4.22. Jerky can be made economically from very lean ground beef rolled into a very thin sheet.*

*Figure 4.23. The ground meat needs to be rolled to a uniform thickness of ⅛ to ¼ inch. Half a pound of meat makes a rectangle approximately 9 x 12 inches.*

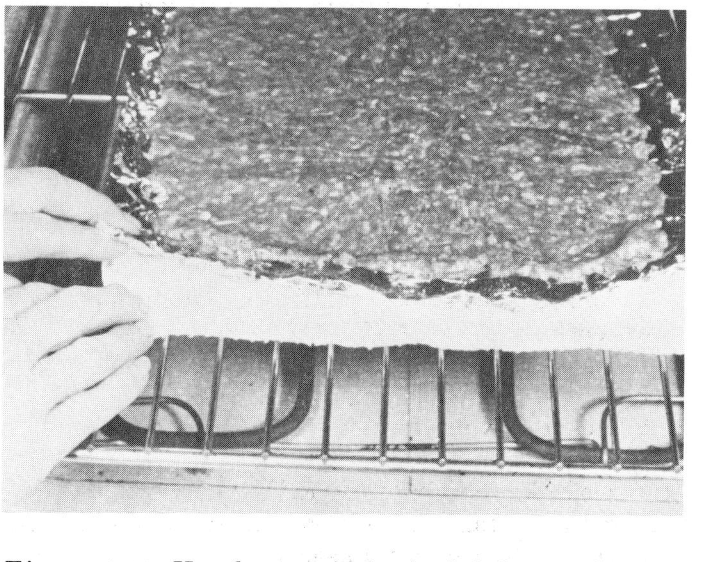

*Figure 4.24. Hamburger jerky is dried on a base of paper toweling and aluminum foil. The toweling will absorb the drippings and promote drying. As the toweling becomes moist, it should be replaced with fresh towels, and the meat should be turned occasionally.*

## Other Types of Jerky

Jerky may be made using other meats and fish, too. Liver and heart are organ meats which can be used satisfactorily because of their rather low fat content. Cooked meats of all kinds also may be sliced thin or cubed and dried for later use in soups and casseroles. Pork and wild game should always be cooked prior to drying; this measure avoids the risk of trichinosis.

Fresh fish and seafood may be cooked prior to drying, if desired. If fish is to be dried without cooking, it should be salted and refrigerated for at least 12 hours prior to drying. Salting is done using two tablespoons of salt per pound of fish. Special flavorings may be introduced by using onion salt, garlic salt, or soy sauce as the salt source.

# USING DRIED FOODS

Dried foods may be used in many different ways. Many of them are excellent when eaten in their dried form. The dried foods are well suited to use as snack items. Slices of dried fruits or dried vegetables are pleasing to eat in their dried form. However, dried foods can assume additional roles in the diet when they are incorporated with other ingredients to make a variety of products. Some suggestions for rehydration and use are provided in this section.

## Vegetables

Dried tomatoes and other vegetables are well suited for inclusion in soups, stews, and other casseroles. They may be rehydrated before being used in these dishes, but the liquid in the recipe itself often is sufficient to allow rehydration during the time required for cooking the other ingredients in the product. These dried vegetables also can be incorporated by crushing the dried food into a powder to facilitate uniform distribution throughout a dish. This is particularly effective when the cooking time of the recipe is short and the ingredient is a flavoring item, such as dried onion.

### Calico Dip

½ cup dried mixed vegetables (onions, red and green peppers, chili peppers, carrots, celery, cabbage, zucchini, tomatoes, pimiento, mushrooms, or others)
1 package (1½ oz.) stroganoff mix or chicken gravy mix)
Hot water to cover
1 cup sour cream
8 ounces cream cheese, softened
Seasonings to taste (seasoned salt, seasoned pepper, garlic powder, or other seasonings)

1. Mix dried vegetables together in a small bowl.
2. Add enough hot water to cover vegetables, and cover tightly.
3. Let stand for one hour to rehydrate.
4. Stir in sour cream and cream cheese.
5. Blend in seasonings to taste.
6. Refrigerate (covered) for at least 30 minutes to allow flavors to blend.
7. Serve as a dip for chips, crackers, raw vegetables, or with dried zucchini chips. This recipe also is appropriate as a sandwich filling with meat or in an open-faced broiled sandwich. When thinned, this dip is a flavorful salad dressing particularly well suited for use with a mixed green salad.

### Creamed Dried Corn

1 cup dried corn (Pride of Iowa)
Water to cover
Cream
Salt to taste

1. Cover corn with water and soak, covered, in the refrigerator overnight.
2. Simmer an hour in a covered saucepan, stirring occasionally.
3. Add cream and salt to taste.
4. Heat to serving. Makes 6 servings.

### Patchwork Meat Loaf

2 pounds ground beef
2 eggs
½ cup dried vegetables (chopped)
1 package mushroom gravy mix
1 teaspoon seasoned salt
Other seasonings, as desired
1½ cups milk (or other liquid)

1. Beat the egg in a medium-sized bowl.
2. Stir in the milk, dried vegetables, gravy mix, and seasonings until the mix dissolves completely.
3. Add the meat and mix until all ingredients are well blended.
4. Transfer the mixture into a 9 x 5 x 3 inch loaf pan and shape into a loaf.
5. Bake at 350°F. for 50 minutes until done.
6. Let stand 5 minutes; then slice and serve.

### Indian Corn Pudding

¾ cup dried corn
2 cups light cream (half and half)
2 eggs, slightly beaten
2 tablespoons chopped green onions ( 1 teaspoon dried chopped onion)
2 tablespoons melted margarine
1 tablespoon sugar
1 teaspoon salt
⅛ teaspoon pepper

1. Heat the cream and add to the rinsed dried corn.
2. Cover and let soak about 4 hours.
3. Combine the remaining ingredients with the corn-cream mixture.
4. Pour into greased 1-quart casserole.

5. Bake, uncovered, in a 325°F. oven for 35 minutes or until a knife inserted in the center comes out clean. Serves 6.

*Note:* Makes 2 quarts.

**Tomato Rice Soup**

2 cups boiling water
1 cup dried mixed vegetables
1 package of brown gravy mix
1 pound ground beef
1 can (1 lb. 12 oz.) tomatoes
2½ cups water
1 teaspoon salt (to taste)
⅓ cup long grain polished rice

1. Pour boiling water over dried mixed vegetables and gravy mix in bowl.
2. Cover bowl with plastic wrap and let stand 30 minutes to rehydrate vegetables.
3. Saute ground beef and remove excess fat.
4. Add the undrained vegetables, the tomatoes, water, and salt.
5. Heat to a boil; then cover and reduce heat to a simmer.
6. After simmering 35 minutes, add the rice. Cover and continue simmering another 15 minutes just until rice is tender. Makes 2 quarts.

**Zucchini Italiano**

1 cup water
1 cup dried zucchini (sliced or chopped)
¼ cup dried onions (sliced or chopped)
1 teaspoon garlic salt
1 teaspoon mixed Italian seasonings (oregano, thyme, or others as desired)
1 can (1 lb.) or 1 pint stewed tomatoes
2 tablespoons margarine

1. Place zucchini, water, and onion in a saucepan. Bring to a boil.
2. Cover the pan and let stand 20 minutes.
3. Add tomatoes, garlic salt, other seasonings, and margarine.
4. Cover and simmer for 15 minutes. Adjust seasonings to suit taste, if desired.
5. If a thicker vegetable is desired, mix 2 teaspoons cornstarch thoroughly with 1 tablespoon cold water. Gradually add this mixture to the zucchini while stirring. Stir constantly while heating to a boil. Another variation is provided by adding a slice or two of white bread, torn in small pieces, in place of the cornstarch thickening.

**Peppy Green Beans**

12 ounces tomato juice or vegetable juice cocktail
1 cup dried green beans
1 teaspoon salt
2 teaspoons horseradish
Grated cheese (Parmesan, cheddar, or other favorite)

1. Heat tomato juice in a small saucepan.
2. Remove from heat and add the dried beans to rehydrate.
3. Cover saucepan and let stand 1 hour.
4. Stir in the salt and horseradish.
5. Pour mixture into greased 1-quart casserole.
6. Sprinkle top with grated cheese.
7. Bake, uncovered, for 30 minutes at 350°F.

## Fruits

**Country Creamy Apple-Raisin Pie**

**Filling**

1 cup dried apples (snipped)
½ cup raisins
1 egg
¾ cup sugar
2 tablespoons flour
1 cup sour cream
½ teaspoon vanilla
¼ teaspoon salt

**Crust**

1½ cups flour
½ teaspoon salt
½ teaspoon sugar
½ cup oil
3 tablespoons milk

**Topping**

⅓ cup sugar
⅓ cup flour
1 teaspoon cinnamon
¼ cup butter
½ cup chopped walnuts or pecans

1. Cover dried apples and raisins with boiling water and let stand one hour.
2. While fruit is rehydrating, prepare the crust by first mixing together the flour, salt, and sugar in a pie plate.
3. Whip the oil and milk together vigorously with a fork in the measuring cup.
4. Add liquid to the flour mixture and mix with a fork until well blended.
5. Press the dough to a uniform thickness across the bottom and up the sides of the pie plate. Flute edge of dough or make other attractive edge.
6. Bake crust 7 minutes in oven preheated to 425°F.
7. Crumble the topping ingredients together, using a pastry blender. Set aside.
8. Beat the egg and remaining filling ingredients together thoroughly.
9. Drain the fruit and add the apples and raisins to the filling.
10. Pour the filling into the hot crust.
11. Bake immediately in a 400°F. oven for 25 to 30 minutes until almost set.
12. Lower oven setting to 350°F. while sprinkling the topping over the pie.
13. Bake 10 minutes at 350°F. Serve warm or cold.

**Dried Fruit Gelatin**

1 package (3-oz.) fruit-flavored gelatin
2¼ cups boiling water
½ cup dried fruit

1. Place the powdered gelatin and the snipped or sliced dried fruit in a bowl.
2. Pour the boiling water over the fruit and gelatin while stirring vigorously.
3. Continue stirring until all of the gelatin is dissolved.
4. Cover bowl tightly and let stand at room temperature for 30 minutes to rehydrate fruit.
5. Refrigerate the gelatin until it sets. This may be done in the bowl or in molds. Serves 4.

**Compote Cobbler**

2 cups dried fruit (about 8 oz.)
3 cups apple cider or water
⅓ cup sugar
1 tablespoon cornstarch
2 teaspoons grated lemon rind
2 tablespoons lemon juice
1 package (9 oz.) white or yellow cake mix
¼ cup margarine or butter, melted
Chopped or slivered nuts, as desired

1. Cut fruit into uniformly large pieces.
2. Place fruit and liquid in a small, covered saucepan.
3. Heat to a boil, reduce heat, and simmer 15 minutes or until tender.
4. Thoroughly blend sugar, cornstarch, and lemon rind and stir into hot fruit mixture.
5. Bring to a boil, stirring constantly.
6. Stir in the lemon juice before pouring mixture into greased, metal 8 x 8 inch baking pan.
7. Sprinkle dry (powdered) cake mix over the hot fruit.
8. Pour melted margarine over the dry cake mix, and sprinkle generously with nuts.
9. Bake 45 to 50 minutes at 350°F. until pleasingly browned.
10. Serve warm with whipped cream or ice cream. Serves 6 to 9.

**Apricot Cookies**

1 cup chopped dried apricots or apricot leather
¾ cup boiling water
2 cups brown sugar
½ cup shortening
½ cup margarine or butter
3 eggs
1 teaspoon baking soda
1 teaspoon vanilla
3 cups flour
1 teaspoon salt
½ cup chopped nuts

1. Pour boiling water over the dried apricots and allow to cool.
2. Cream shortening, margarine, and sugar together.
3. Beat in the eggs, one at a time.
4. Pour off ⅓ cup of the water from the cooled apricots, and discard any water still remaining.
5. Add the ⅓ cup liquid from the apricots, the vanilla, and the soda to the creamed mixture.
6. Stir in the drained apricots.
7. Toss together the flour, salt, and nuts; stir into the creamed apricot mixture and stir gently until blended.
8. Drop by teaspoonsful onto a greased cookie sheet.
9. Bake until brown (about 15 minutes) in an oven preheated to 350°F.
10. Ice with Browned Butter Frosting (below).

**Browned Butter Frosting**

¼ cup butter or margarine
2 cups powdered sugar
Cream, as necessary for spreading consistency

1. Brown the butter over medium heat in a small saucepan until color is rich and golden.
2. Stir in the sifted powdered sugar.
3. Add cream as needed to achieve spreading consistency.
4. Ice Apricot Cookies (above) with this frosting while the cookies are still warm.

## Jerky

**Hamburger Jerky Pizza**

Flour tortillas
Pizza sauce
Hamburger jerky
Grated cheese (Mozzarella, Parmesan, Romano)
Sliced black olives and other pizza toppings, as desired

1. Place flour tortillas on baking sheets to serve as the base or dough for pizza.
2. Spread thickly with pizza sauce.
3. Sprinkle generously with crumbled jerky, the assorted cheeses, and other topping ingredients, as desired.
4. Bake in 425°F oven until ingredients are heated and cheeses are melted.
5. Serve hot as soon as removed from oven.

# Chapter 5. Canning

## INTRODUCTION

Canning is a time-honored method of preserving foods using heat to destroy bacteria, yeasts, and molds which would cause unprocessed foods to spoil. For canned foods to be safe, it is essential that processing times and temperatures be monitored carefully and that a good seal be achieved. This is true both for home- and commercially-processed canned foods. Unless microorganisms in the food prior to processing are destroyed by heating and unless subsequent contamination is prevented by a tight seal, spoilage will occur in the canned product during storage on the shelf. The potential problems in canned foods represent not only economic loss due to spoilage of food which could have been utilized, but also health hazards, the most serious of which can culminate in death of the unfortunate consumer. Such problems need not happen. Canning is an excellent means of preserving food, but knowledge of the appropriate procedures and diligent application of the principles are essential ingredients in this method of food preservation. For canning to be successful, some key equipment is required. Many additional items are available, but are not vital to achieving safely-preserved products.

### Reasons for Canning

Canning can provide food of excellent quality at lower cost than canned foods purchased in the store, or canning can actually cause foods to be more costly than commercially canned foods. The relative costs of the two approaches are determined by the amount of food being processed in the home canning equipment, the safety of the food being canned at home, and the cost of the food being utilized for home canning. To illustrate, a jar of home canned food would be extremely expensive if a person bought a pressure canner and jars plus closures, then decided after canning a couple of quarts of food that this was not a method of food preservation which met the family's time and energy limitations. The cost of the equipment would need to be distributed over very little food. Conversely, repeated use of canning equipment over many years brings the cost of the equipment to a miniscule level when considered on a per jar basis.

Canning also can be expensive if jar breakage or spoiled food is a part of the total experience. With correct use of jars designed for canning and attention to proper processing techniques and times, losses due to breakage and spoilage should be eliminated. In this case, the loss factor is no different for home canning than it is for commercial canning.

The factor probably most influential in determining advisability of home canning in relation to cost is the price which must be paid for the food being canned at home. If the food must be purchased at the grocery store, it may be sufficiently expensive to make home canned foods more expensive than their commercial counterparts. However, when seasonal values are available, home canning can represent a saving. The greatest savings are made when fresh produce is available from the garden or from fruit trees. Indeed, canning may be an essential means of avoiding food waste when a garden is bearing abundantly and contributing food beyond the amount which can be consumed at the time.

Yet another motivation for home canning is the desire to have fruits and vegetables of the quality provided by home canning. By doing it at home, it is possible to control the size and shape of pieces and to modify sugar and salt content to satisfy individual preferences. For persons on special diets, home canning affords the opportunity to prepare food with known sweeteners or other additives. The sugar content, for example, can be identified and controlled readily when the canning is done at home. Persons wishing to avoid the possibility of additives in commercially canned products may prefer to can their foods at home.

Home canning is a practical hobby or an outlet for creativity for some people. Although the food is useful and enjoyed by them later, there is a genuine pleasure in the accomplishment of canning beautiful jars of food. This satisfaction and feeling of accomplishment cannot be duplicated by buying commercially canned foods. The pleasure of creative food preservation also can be shared with others when specialty items are presented to friends as unique and very personal gifts.

# EQUIPMENT

## Containers

Glass jars, ranging in size from ½ pint to ½ gallon are available for home canning. These are manufactured carefully to meet thermal shock requirements encountered in home canning. In order to avoid the possible breakage of jars during heat processing and the consequent loss of food, it is wise to use jars designed for canning rather than using other types of jars (even if a food product was purchased in the jar previously). Even when canning jars are being used, it is important to examine each jar individually for any flaws in the jar itself and for any imperfections in the rim.

Jars for canning vary slightly in their design and proportions, depending on the manufacturer. In addition to choices between manufacturers, there are choices between wide-mouthed and regular jars. Either can be used for most canning, although there may be a few instances when the size of the piece desired in the canned product may dictate the need for the wide-mouthed jar. The size of jar chosen is dependent upon the amount of food which is convenient to process at the time, the amount of food which is suited to serving the family after the canned food has been opened, and also the size of the processing equipment available for canning. To illustrate, if many quarts of tomatoes were awaiting processing, it would be more efficient and practical to can in quart jars rather than in pints. A person canning for a family of four or more would be able to utilize an opened quart jar of tomatoes before they would spoil. However, a person living alone probably would find a pint of tomatoes more convenient than a quart. The third determinant, the size of the processing container, is of particular interest to people who may be canning for the first time. It is wise to determine the likelihood of continuing to can in the future before purchasing special equipment. Equipment large enough to process pints safely is more likely to be available in existing kitchen equipment than is the larger equipment required for quarts.

*Figure 5.1. Canning jars are available in various sizes and shapes. Different closures also are sold for home canning. Because of the thermal shock to jars used in canning, it is important to use these special containers rather than using jars which formerly contained commercially-processed foods.*

Tin cans also sometimes are used in home canning. These have the advantage of being less breakable than glass canning jars, but they have the disadvantage of requiring a special piece of equipment, the can sealer, for closing them properly. The can sizes ordinarily used are the No. 2 can, which holds 2½ cups, and the No. 2½, holding about 3½ cups. These cans may be purchased with a plain tin lining, with a C-enamel lining, or with an R-enamel lining. The C-enamel, sometimes termed the corn-enamel, is especially designed to meet the requirements of corn and hominy so that there will be no interaction between the can and its contents. The R-enamel, also called sanitary or standard enamel, is tailored to resist the corrosive qualities of red berries, red and black cherries, beets, plums, rhubarb, winter squash, and pumpkin. All foods not specifically mentioned above can be canned satisfactorily in the plain tin cans. Lids for sealing the cans should be the correct size for the can and should have perfect gaskets. Lids are retained in their paper packing until being used and should merely be rinsed (not washed) prior to use. The gaskets are quite susceptible to damage, which will cause cans to have sealing failures.

## Closures

Lids designed especially for sealing the canning jars are essential to safe canning. The styles of closures have changed over the years, but the type used most commonly today is of two-part construction. The actual cover is a flexible metal, undercoated with plastic to prevent interaction of food with the metal of the lid. A ring of special sealing compound encircles the underside of the lid. This compound is responsible for the seal which must be formed for canning to be effective as a means of preserving foods. Therefore, each lid needs to be examined to be certain that the sealing compound is distributed uniformly all around the lid and that there are no scratches. Lack of compound or scratches can cause the lids to fail to seal. These closures are designed for a single use. New lids must be used each time. The other part of these two-piece closures is reusable. This is the metal band used to hold the lid in position on the jar until the seal is accomplished. As long as these bands are not corroded or bent, they can be reused.

Recently a reusable two-piece lid has been developed. The lid actually consists of a flexible plastic which adheres to a washable rubber-like gasket. A metal band is used in the same role as that discussed above. However, it is necessary to tighten this ring within a half hour after processing is completed so that the rubber gasket is held tightly to the jar. The processed jars are stored with the metal bands tightly secured.

Zinc lids with porcelain linings have been used for many years in conjunction with a rubber gasket which fits tightly on the ridge at the shoulder of the canning jar. These closures work satisfactorily, but are not used frequently today.

A one-piece metal lid also is available in some markets. Kuhn and Hamilton (1976) found this closure to be very satisfactory. These same workers tested other closures also. The current interest in home canning seemingly has prompted development of several designs of closures. These new closures should be evaluated carefully to be certain that a successful seal will be achieved.

## Processors

### Water bath canner

A water bath canner (also referred to as a boiling-water-bath canner) is simply a large, deep kettle equipped with a rack for holding the jars and a tight-fitting cover. Canners designed specifically for canning fruits can be purchased, but any kettle with the appropriate rack and lid can be used. Traditionally, kettles deep enough to allow jars to stand upright on the rack and be covered with at least an inch of actively boiling water have served as water bath canners. Ideally, the canner should be deep enough to permit four inches of water above the tops of the jars plus ample space for the boiling action needed during processing. This large a kettle, when filled with jars of food and the boiling water, is very heavy and awkward to handle. In addition to the possible safety hazard in handling such a kettle, considerable energy is needed to heat so much water to boiling. With current energy concerns, there is reason to consider the safety and advisability of modifying traditional techniques.

The need for such a large amount of water for water bath canning was questioned by Harris and Davis (1976). These researchers surveyed the types of home canning vessels available in their area of Alabama and found that none of the canners de-

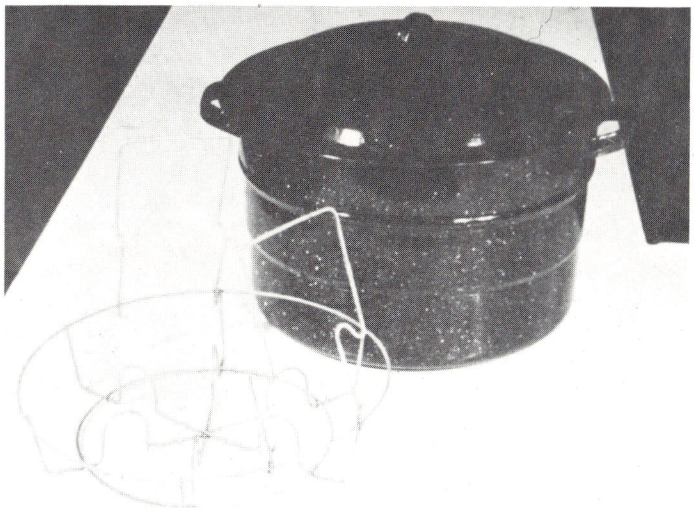

*Figure 5.2. A water bath canner is simply a large kettle equipped with a rack and a lid. This is used for canning fruits and any high-acid foods.*

signed for atmospheric pressure processing (water bath canners) and only two of the pressure vessels were deep enough to meet the recommended depth of four inches of boiling water above the top of the jars being processed. In view of the questionable design of the existing containers, there is a great likelihood that most home canned fruits are not being canned with the recommended level of water. This raises, then, the question of the safety of these products.

To determine the suitability of using lesser amounts of water, Harris and Davis (1976) determined the final temperature reached in tomato juice canned in atmospheric and loosely closed pressure canners with the depth of water in each vessel ranging from one to eight inches. The juice was found to heat as fast when the water depth in the canner was 2.5 inches as when the water depth ranged as high as one inch above the tops of the jars. Not only does this finding make it practical to use available equipment (with its slightly shorter height than is recommended), but it also suggests the possibility of conserving heat energy and time. In the study, time and heat energy required for preheating the water to boiling were reduced to about a fourth of the time and heat energy needed when the water level was an inch above the tops of the jars. Of course, these savings also mean that the kitchen will not get as hot during canning if the lower water level is used as it will when the very high, recommended level of water is used.

### Pressure canner

Vegetables and other low-acid foods need to be processed at temperatures higher than can be reached with water bath (atmospheric pressure) canners. The problem is solved by use of the pressure canner. Pressure canners are cast from thick metal and are equipped with a rack, a very tight, locking cover, and a pressure gauge. The cover is designed with a petcock for allowing exhaustion of air and steam from the canner prior to applying pressure to

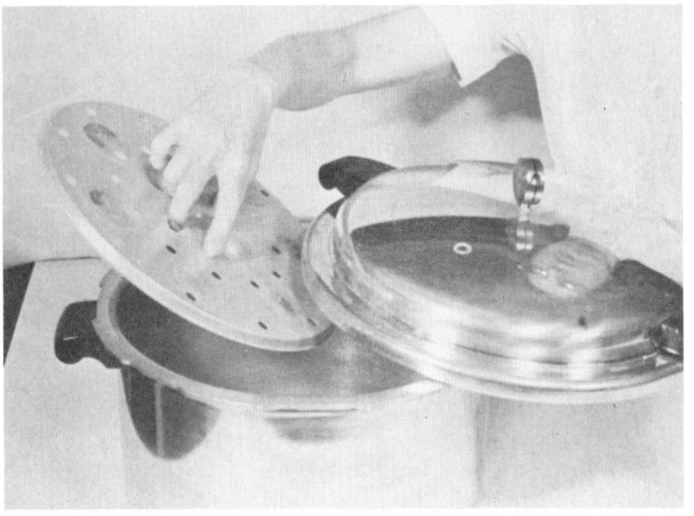

*Figure 5.3. Pressure canners are heavy metal vessels equipped with a sturdy rack for the jars, a very tightly-fitted lid, a pressure gauge, and a safety valve.*

the vessel. It also is equipped with a safety-valve opening to guard against unsafe pressure levels. The direction manual for the pressure canner should be available whenever the canner is being operated so that the specific safety precautions for the vessel can be observed. A pressure canner is a particularly appropriate choice of equipment for people doing a great deal of canning because it can be used with pressure for low-acid foods and without pressurizing for fruits, thus eliminating the need for a boiling-water-bath canner. However, this equipment represents a large investment if it is used but rarely.

*Figure 5.4. Pressure canners are deep enough to hold quart jars when the jars are standing on the rack in the canner.*

**Pressure saucepan**

A pressure saucepan can be used for canning half pint and pint jars, but it may not be deep enough to accommodate quart jars. As with the pressure canner, a pressure saucepan must be operated with a rack to hold the jars off the bottom of the vessel. The lid is similar in its features to the lid of a pressure canner. The booklet accompanying the pressure saucepan should be consulted for guidance on operation of the saucepan and for processing times.

Pressure canners and pressure saucepans must be maintained in excellent operating condition. If equipped with a dial gauge, it is necessary to have the accuracy of the indicating dial checked annually. County Home Extension Agents can provide information on the local location where this can be done. The other precaution is to be certain that the petcock is not clogged. This can be tested quickly by simply inserting a small skewer through the valve opening. If the canners and saucepans are equipped with weighted, rather than dial gauges, there is no need for checking other than being sure that the gauge is clean.

*Figure 5.5. Pressure may be measured or controlled by either a dial gauge (left) or a weighted gauge (right).*

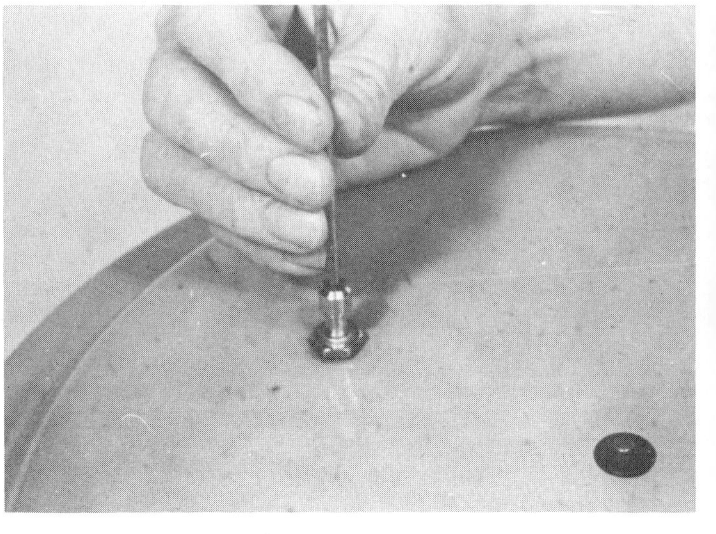

*Figure 5.6. A skewer can be inserted through the petcock to be certain that nothing is blocking the channel.*

# FOOD FOR CANNING

## Selection of Food

Foods for canning should be of excellent quality and fully ripe, but not too ripe. Fruits should still be just slightly firm, and vegetables should be young and tender. Meats, poultry, and fish to be canned also need to be of excellent quality. Freshness of any food to be canned is of great importance, for the quality of the canned food can never be better than the food placed in the jar. Ideally, produce is harvested just prior to canning; meats, poultry, and fish are through rigor, but still very fresh, with firm flesh. By selecting foods carefully, the canned foods also will be of optimum quality.

Although the quantity of canned food which will result from a given weight of fresh produce is somewhat variable, it is helpful to have at least an approximate yield which can be anticipated. Tables 5.1 and 5.2 provide useful guidelines to yields.

Table 5.1. Amount of fresh fruit required to yield 1 quart of canned fruit

| Fruit | Pounds |
|---|---|
| Apples | 2½ to 3 |
| Berries (except strawberries) | 1½ to 3 (1 to 2 quarts) |
| Peaches | 2 to 3 |
| Pears | 2 to 3 |
| Plums | 1½ to 2½ |
| Tomatoes | 2½ to 3½ |

Table 5.2. Amount of fresh vegetables required to yield 1 quart of canned vegetable

| Vegetable | Pounds |
|---|---|
| Asparagus | 2½ to 4½ |
| Beans, lima (in pods) | 3 to 5 |
| Beans, snap | 1½ to 2½ |
| Beets, without tops | 2 to 3½ |
| Carrots, without tops | 2 to 3 |
| Corn, sweet (in husks) | 3 to 6 |
| Okra | 1½ |
| Peas, green (in pods) | 3 to 6 |
| Pumpkin or winter squash | 1½ to 3 |
| Spinach, other greens | 2 to 6 |
| Squash, summer | 2 to 4 |
| Sweet potatoes | 2 to 3 |

## Preparation of Food

Foods for canning need to be washed thoroughly to insure that all dirt and surface contaminants have been removed. This is an essential step in obtaining food as free of bacteria, yeasts, and molds as possible. Use of well cleaned food helps to insure a satisfactory and safe product following processing. A vegetable brush or other rather abrasive material will facilitate removal of dirt and its accompanying microorganisms. Thorough washing of greens to be canned can best be done by agitating the greens in a sink filled with water several times and then rinsing the sink well between washings to remove the sediment which collects in the bottom of the sink. The process must be repeated until there is no dirt or sand remaining.

Vegetables for canning need to be trimmed with a paring knife to remove areas of heavy cellulose and any blemishes. A few vegetables, such as beets, tomatoes, and potatoes need to be peeled. The skins from potatoes are pared with a potato peeler. Beets need to be cooked until tender with their jackets on; then they can be peeled simply by pulling off the skin. Tomatoes can be peeled easily following a 30-second blanching. The ends of most vegetables also need to be trimmed off, and the kernels of corn should be cut from the cob with a sharp knife.

Fruits to be canned also need to be washed and cleaned well prior to processing. Berries should be hulled or capped and any pieces of leaves removed. Grapes are plucked from the stems. Pears and apples are pared, either sliced or halved or quartered, and cored. Peaches are dipped in boiling water to loosen their skins so that the skins can be pulled off with a knife. The pits are removed from the pared peaches, and the fruit is ready for packing either as halves or slices. Apricots are pitted, but not peeled. Cherries have their stems removed, and usually they are pitted.

*Figure 5.7. Tomatoes (or any other food being prepared for canning) need to be meticulously clean before being packed for processing.*

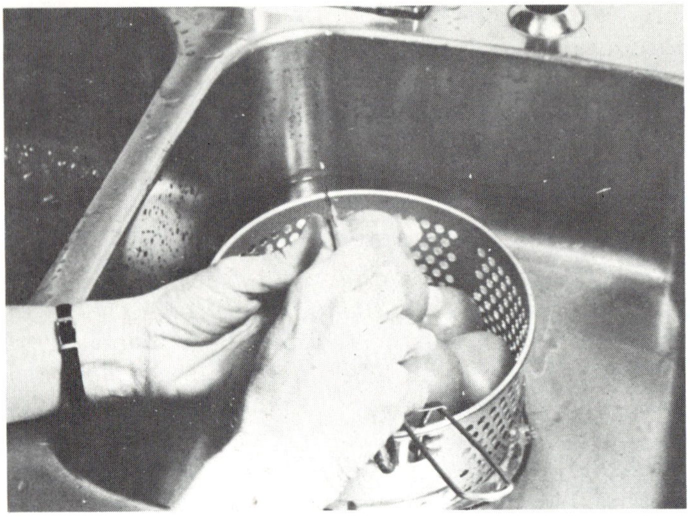

*Figure 5.8. After a brief blanching and quick immersion in ice water, tomatoes can be peeled easily in preparation for packing for canning.*

Some fruits, such as peaches, pears, and apples, discolor very quickly. When this occurs, the appearance of the canned product will not be as attractive as it would have been had the color been retained. Browning of fruits being prepared for canning can be avoided by placing the pared fruit slices in any of the following: (1) a mild salt solution made by dissolving 1½ teaspoons salt per quart of water; (2) dilute lemon juice (2 tablespoons per cup of water); or (3) fruit juice, such as pineapple juice. Ascorbic acid (either 50 milligrams in the bottom of each quart jar or ½ teaspoon of the crystalline powder in each quart of sugar syrup used in packing the jars of fruit) can be used to inhibit browning in the canned fruit.

Meats, poultry, and fish also may be canned very successfully. For economy in packing jars, removal of bones and excess fat is wise. This procedure saves considerable space, particularly when chicken is being canned. Ordinarily these foods are cut into serving size pieces or into smaller pieces for use in stews and casseroles. Poultry may be packed

*Figure 5.9. Browning of pears being prepared for canning can be prevented by immediately immersing the pared, sliced fruit in an ascorbic acid solution.*

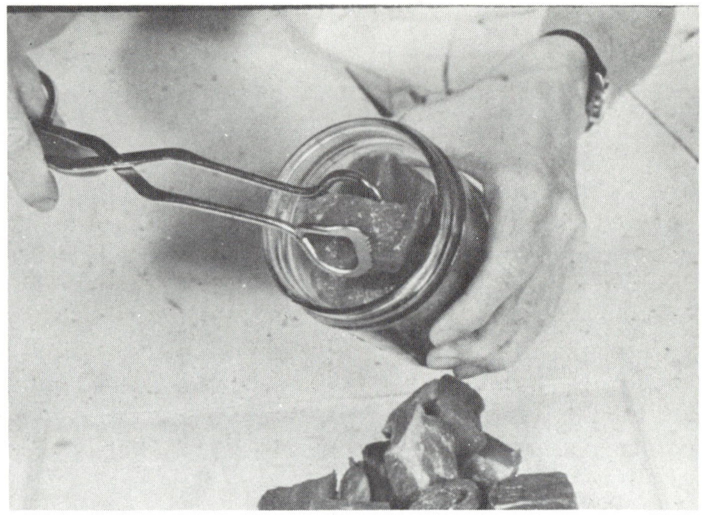

*Figure 5.10. Valuable space can be saved in canning jars by removing bones and fat and cutting meats in cubes.*

raw or precooked. Pork also may be used raw or cooked. Beef, lamb, and veal often are packed without precooking, but cooked meats can be packed. If fat content is fairly high in meats, as is true with some pork and sausage, precooking may be preferred because this procedure makes it possible to remove more of the fat prior to processing the canned food. Fish are packed for canning either raw or cooked.

## Method of Packing

### Hot Pack

Hot pack is defined as the filling of canning jars with hot, pre-cooked food. This method of filling jars often is done with meat, poultry, fish, pickles, relishes, and some legume recipes. The advantage of hot packing vegetables prior to packing is that the food will shrink somewhat less during processing. However, the added cooking which the vegetables receive when they are cooked prior to heat process-

*Figure 5.11. Poultry is canned using the hot pack method.*

ing may cause the food to be softer and less desirable. Consequently, hot packing is not done frequently. When it is done, it is wise to minimize nutrient losses by using the water used for pre-cooking as the liquid for filling the jar.

After the food has been packed as efficiently as possible in the jar, boiling water (or preferably the pre-cooking liquid) is poured to leave just half an inch of headspace. Exceptions to the half inch of headspace are: applesauce (¼ inch headspace), baked beans (1 inch), lima beans (¾ inch for pints and 1¼ inches for quarts), corn (1 inch), peas (1 inch), and squash (1 inch).

**Cold pack**

The cold pack method is used for most foods. In this method it is important to pack the food together tightly because some volume will be lost during processing due to shrinkage of the vegetable or fruit. Cold packing is done simply by placing the pieces of food in the jar and then pouring boiling water (for vegetables), sugar syrup (for fruits), or broth (for meats, poultry, or fish) over the food in the jar until the proper amount of headspace remains. With the exceptions of the starchy foods mentioned in the hot pack method, the headspace should be half an inch.

When a sugar syrup is used for canning fruits, the syrup can be prepared as a thin, medium, or heavy syrup to suit individual preferences. The sugar syrup helps to retain the firmness of the fruit and to modify the flavor. To make five cups of thin syrup, two cups of sugar are heated together with four cups of water until the sugar dissolves. For preparing 5½ cups medium syrup, three cups of sugar and four cups of water are used; for 6½ cups heavy syrup, 4½ cups of sugar are heated with four cups of water until dissolved. Honey can be used, if desired, in place of sugar. The suggested level of honey for preparing a syrup for fruits is one

*Figure 5.12. Vegetables and fruits frequently are packed by the cold pack method, and then liquid is added to leave the appropriate amount of headspace.*

*Figure 5.13. Pears are packed using the cold pack method, and then a sugar syrup of the desired sweetness is added.*

Table 5.3. Sugar syrups for canning fruits

| Type | Composition of syrup | Yield (cups) |
|---|---|---|
| Thin | 2 c. sugar/4 c. water | 5 |
| Medium | 3 c. sugar/4 c. water | 5½ |
| Heavy | 4½ c. sugar/4 c. water | 6½ |
| Honey | 1½ c. honey/4 c. water | 5½ |
| Thin honey | ¾ c. honey, ¾ c. sugar/4 c. water | 6 |
| Corn syrup | | |
| Thin | 1 c. corn syrup, 1 c. sugar, 4 c. water | 5½ |
| Thin | 1 c. corn syrup, 1 c. sugar/4 c. water | 5½ |
| Medium | 1½ c. corn syrup, 1½ c. sugar/4 c. water | 6 |
| Heavy | 2 c. corn syrup, 2½ c. sugar/4 c. water | 7½ |

to one and a half cups of honey and four cups of boiling water. If the flavor of honey desired is to be more subtle, half the amount of honey suggested can be replaced with sugar. Corn syrup also can be used up to replacing half the sugar suggested for the sugar syrups above. If more than half of the sugar is replaced by corn syrup, the flavor may be discernible. Brown sugar, molasses, or other strong-flavored or dark-colored sweeteners are not recommended for use in canning fruits because of their detrimental effects on color and flavor. If desired, fruits can be canned with no sugar or other sweetener added. Since the sugar content in canned fruits is too low to serve as a preservative, the processing times are the same with or without added sugar.

Some persons, particularly diabetics, may wish to can fruits using artificial sweeteners. The directions for use offered by the manufacturer provide a guide to making such substitutions. The processing times for the canned fruit will not be altered.

Salt can be added to vegetables and to meats, if desired. When added to vegetables, the suggested amount of salt to use is ½ teaspoon to pints and a teaspoon to quarts. The salt is added after the

*Figure 5.14. Salt is added to meat being cold packed for canning.*

*Figure 5.16. Closures for the jars need to be prepared for use according to the manufacturer's directions.*

vegetable has been placed in the jar. The boiling water then is poured over the salt and the vegetable, leaving an inch for headspace. If salt is added to meat to be canned, the amount used is the same as for vegetables (½ teaspoon for pints and 1 teaspoon for quarts).

It is important to avoid air pockets in the jars of food. Any trapped air can be relieved easily by gently inserting a narrow rubber or plastic spatula down the side of the jar. Air pockets can be noted by looking through the side of the jar. This step helps to facilitate exhaustion of the jar.

## Sealing the Jars

The closures for the jars should be prepared for use according to the manufacturer's directions. For the traditional two-part closures, this means washing and rinsing the lids carefully and, in some instances, heating or boiling the lids and rings the length of time specified by the manufacturer. The treatment is dictated by the sealing compound used in the lid. If porcelain-lined zinc lids are being used, new rubber rings are needed, and these must be washed well in soapy water and rinsed thoroughly.

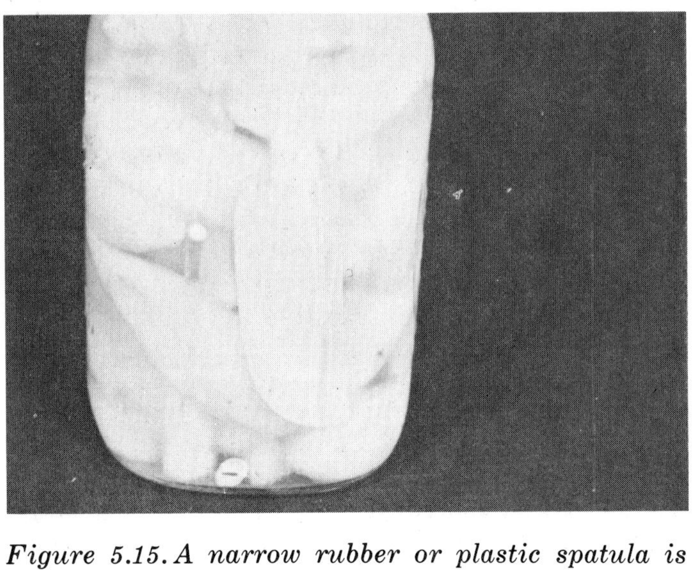

*Figure 5.15. A narrow rubber or plastic spatula is inserted carefully to help release any air bubbles which may be trapped in the jar of food.*

### Two-piece lids

The filled jar must have a perfectly clean rim for a seal to form. Therefore, the rim of the jar must be wiped perfectly clean before the flat lid is applied with its sealing compound ring adjacent to the rim of the jar. The use of tongs for transporting the lid from the scalding water to the jar is a practical means of helping to keep the food and its package as clean as possible. The metal band then is screwed tightly to the jar by hand. It is important to do this part of the operation firmly, but not so firmly that the rim of the jar is forced through the sealing compound. Too tight a closure will damage the sealing ring and it will not be possible to get the necessary seal on the jar. This hand tightening will give just the right amount of pressure on the closure to enable the lid to function according to its design during processing, that is to let the air in the jar escape.

### Porcelain-lined zinc caps

To begin to seal a porcelain-lined zinc cap, it first is necessary to place the wet rubber ring over the mouth of the jar and firmly in contact with the shoulder of the neck of the jar. This step is done before the jar is filled so that there will not be any possibility of food collecting under the ring. After the jar has been filled, the rim of the jar, the exterior of the jar neck, and the rubber ring are wiped clean. The metal cap then is screwed down firmly and turned back immediately a quarter of an inch in preparation for processing.

### Tin Cans

Filled tin cans need to be heated to an interior temperature of 170°F (77°C) or higher before the cans

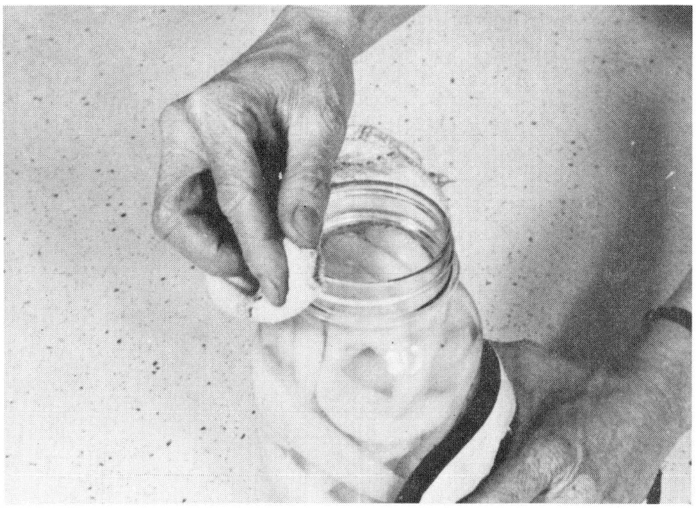

*Figure 5.17. The rim of the jar should be wiped absolutely clean with the aid of a clean, damp cloth. This is necessary so that a good seal can be achieved between the closure and the rim of the jar.*

*Figure 5.18. By using tongs to transfer the prepared closure to the jar, there is little possibility of contamination of the canned food.*

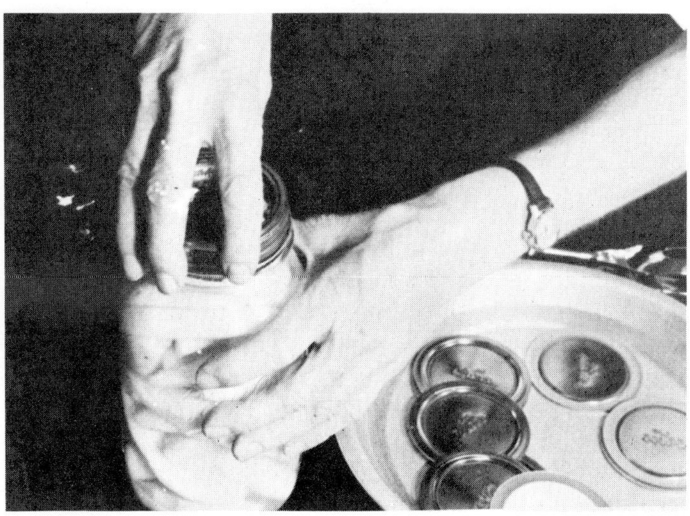

*Figure 5.19. Closures of two-part construction are closed using only the force obtained by the bare hands. Tighter closure will damage the sealing compound in the lid and lead to a faulty seal.*

are sealed. This is necessary to exhaust the air from the cans before the cans are sealed. Tin cans, unlike the glass jars, have an air-tight seal immediately; removal of the air prior to sealing is essential if a good vacuum is to be achieved in the canned product. The different design of the closure in glass jars allows the air to be exhausted from the jars during heat processing so this is not a necessary step when using the glass jars.

If a hot pack is used on foods packed in cans, it may not be necessary to reheat the food before sealing. However, a thermometer should be used to determine that the temperature is above 170°F. If it is not, the cans and the food in them will need to be reheated. Cold pack foods will need to be heated until the thermometer indicates that the minimum temperature (170°F) has been reached in the center of each can. When the correct temperature has been reached, the hot can is removed from the boiling-water-bath (one can at a time), boiling water is poured over the food to reach the desired head space level, the clean lid is placed on the can, and the can is sealed immediately, using the can sealer.

## CANNING METHODS

### Selection of Method

The primary consideration when deciding how to can a food is the matter of food safety. Some of the common food microorganisms are destroyed readily by any of the methods of heat processing. However, some forms are quite resistant to heat destruction. The most resistant problem usually of concern in home canned and commercially-canned foods is the production of toxin by the spores of *Clostridium botulinum*. These spores require a processing temperature of 240°F for a moderately long period of time if they are to be rendered harmless. Since these anaerobic bacteria only reproduce very efficiently in media with a pH above 4.5, it is the low acid foods which require pressurized processing for safety. It should also be noted that it is the pH of the total medium which is of concern. If a mixture of foods (tomatoes, onions, and green pepper, for example) is to be processed together, the pH of the total mixture may dictate the need for pressurized processing even though the tomatoes themselves may have been sufficiently acidic to be safely processed at atmospheric pressure.

Since the need for elevated processing temperatures is determined by the pH of the food, foods have been categorized into either three or four groups, depending upon the authority being cited. Cameron and Esty (1940) suggested that the foods be classified as *low-acid* if they had a pH of 5.9 or higher; as *medium-acid* in the pH range of 4.5 to 5.0; as *acid* in the range of 3.7 to 4.5; and as *high-acid* at pH 3.7 or lower. Stumbo (1973) cites some workers as recommending only three categories: (1) *low-acid* foods, pH above 4.5; (2) *acid* foods, pH 4.0 to 4.5; and (3) *high-acid* foods, pH below

4.0. This classification seems to be fairly adequate since very few microorganisms will grow at a pH as low as 3.7 unless the microorganismal count is extremely high prior to processing. The pH level of 4.5 commonly is accepted as the point separating foods to be processed at atmospheric pressure versus those requiring pressurized processing. Using pH 4.5 as the dividing point, pimientos, okra, pumpkins, carrots, turnips, cabbage, lima beans, corn, and shrimp (in order of decreasing acidity) as well as other vegetables and meats should be processed under pressure. The chart presents a more complete indication of the pH values which can be anticipated for a variety of foods, ranging from the low-acid foods to the high-acid products. Fruits and pickles are the types of foods which are included in the high-acid (low pH) range. Recently there has been considerable research effort directed toward determining the pH levels of various types of tomatoes. This research has been triggered with a rising incidence of unsafe home canned tomatoes. In some instances it appears that the hazard may have been heightened by use of chili peppers and onions, which raised the pH of the mixture. However, there has been considerable variation noted in some of the tomato varieties tested, too (see Table 5.4).

If uncertain about the pH of foods near the critical pH of 4.5, the food can be tested by using pH test paper. A suitable pH range for the paper is pH 3.0 to 5.5. This paper is used by matching the color of the paper after it has been dampened with the food in question with the color samples on the side of the package. If the pH indicates higher than 4.5, processing should be done in a pressure canner.

In an unpublished report in 1976, W. Simpson of U. of I. Research and Extension Center (Parma) reported somewhat higher pH values for the same varieties, as compared with the Eastern Regional Research Center of the Agricultural Research Service report (Table 5.4). He reported Ace at a pH of 5.2, Rutgers at 4.7, and Beefsteak at 5.6. These differences are of significance when one notes that the differences center around the pH range where the decision between atmospheric and pressurized canning must be made. In view of this controversial, yet very important range of acidity, it is recommended that acid be added to tomatoes being canned (regardless of the variety). The types of acids and the amounts which are recommended are as follows:

Lemon juice – 2 teaspoons/quart of tomatoes,
Vinegar – 2 teaspoons/quart of tomatoes,
or
Citric acid, U.S.P. – ½teaspoon/quart of tomatoes.

If pints are being canned, half this amount of the acid selected should be placed in each pint jar.

At the present time, there is some contradictory evidence regarding the potential problems presented by tomatoes of varying acidity. Research at the

Average pH of commercially canned foods

| pH | Food |
|---|---|
| 2.5 | |
| | Plums |
| | Gooseberries |
| | Prunes |
| | Dill pickles, rhubarb, apricots |
| 3.0 | Apples, blackberries |
| | Sour cherries, strawberries |
| | Peaches |
| | Sauerkraut, raspberries |
| | Blueberries |
| | Sweet cherries |
| | Pears |
| 4.0 | |
| | Tomatoes |
| | Pimiento (dry roasted) |
| | Okra |
| 5.0 | Pumpkins, carrots |
| | Pimiento (lye peeled) |
| | Turnips, cabbage |
| | Parsnips, beets, string beans, green peppers |
| | Sweet potatoes, baked beans (tomato sauce) |
| | Spinach |
| | Asparagus, cauliflower |
| | Baked beans (plain sauce) |
| | Red kidney beans |
| | Lima beans |
| | Succotash, meats, poultry |
| 6.0 | |
| | Peas |
| | Corn, hominy, salmon |
| | White fish |
| | Shrimp, wet pack |
| 7.0 | Lye hominy |
| 8.0 | |

Table 5.4. pH of selected varieties of tomatoes

| Variety | Mean pH |
|---|---|
| Ace | 4.47 |
| Beefsteak | 4.23 |
| Burpeanna | 4.37 |
| Early Giant | 4.15 |
| Fireball | 4.50 |
| San Marzano | 4.68 |
| Royal Chico | 4.58 |
| Rutgers | 4.29 |
| Yellow Plum | 4.31 |

U.S.D.A.'s Northern Regional Laboratory suggests that the hazard from botulism in canned tomatoes may be the result of reduced acidity in the jar as a result of mold growing in improperly sealed jars. This laboratory will be continuing its research in this matter in 1977 with the goal of determining the need for adding acid to tomatoes which are to be processed in a water bath. Preliminary findings indicate that this acidification may not be necessary.

## Processing Methods

### Water bath canning

For water bath canning, the hot jars of food are lowered, with the aid of a jar lifter, into a water bath canner containing either hot water (for cold pack fruits) or boiling water for hot pack). The water level needs to be at least 2½ inches, but well may be approximately 6 inches deep when the jars are put into the canner. After all the jars are in position, boiling water usually is added to bring the water level at least an inch above the tops of the jars. It is important that the boiling water not be poured directly on the glass jars because of the risk of breakage.

When the boiling water has been added, the cover is placed on the canner. If a pressure canner is being used for water bath canning, the cover should be positioned securely, but not locked, and the weighted gauge or petcock closure should not be used.

When the water comes to a rolling boil in the water bath canner or steam is escaping in a heavy, steady stream from the open petcock (steam vent) in the pressure canner, timing of processing is begun. The timing required for processing is determined by the type of food being processed and the size of the jars.

The additional time for processing of fruits required because of the effect of altitude on the temperature of boiling water is presented in Table 5.5. Processing times for water bath canning of fruits at sea level are presented in Table 5.6.

Table 5.5. Adjustments in processing times for water bath canning at altitudes

| Altitude (Feet) | Increase in processing time if time in recipe for processing is: 20 minutes or less (Minutes) | More than 20 minutes (Minutes) |
|---|---|---|
| 1,000 | 1 | 2 |
| 2,000 | 2 | 4 |
| 3,000 | 3 | 6 |
| 4,000 | 4 | 8 |
| 5,000 | 5 | 10 |
| 6,000 | 6 | 12 |
| 7,000 | 7 | 14 |
| 8,000 | 8 | 16 |
| 9,000 | 9 | 18 |

*Figure 5.20. A jar lifter is a convenient, safe device for lowering jars into the water bath canner.*

*Figure 5.21. Boiling water is added to the desired depth in the canner. Care must be taken to be certain that the boiling water is not poured directly on any of the jars.*

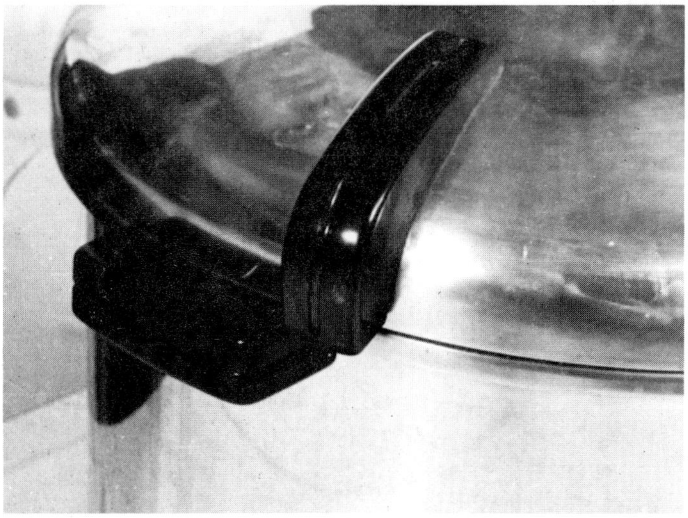

*Figure 5.22. If a pressure canner is being used for water bath canning, the lid is placed on the canner loosely to avoid building up pressure in the canner.*

Table 5.6. Guide to canning fruits in boiling-water bath canner[1,2]

| FRUIT | HOW TO PREPARE | PROCESSING TIME (minutes) Pint | Quart |
|---|---|---|---|
| **Apples** | Pare, cut in halves or quarters, and trim off core. To keep from darkening, dip in 1 gallon water that contains 2 tablespoons each of salt and vinegar. Cook in hot liquid for 2 to 4 minutes, according to the variety. Pack hot. Cover with hot liquid. Seal. | 15 | 15 |
| **Applesauce** | Make with or without sweetening. Add 1 teaspoon lemon juice to each pint jar. Pack boiling hot into hot, scalded jars. Seal. | 10 | 10 |
| **Apricots** | Choose firm, well-colored apricots that are not overripe. To peel, dip in boiling water for 1 minute, then plunge into cold water and peel. Or can apricots without peeling. Leave whole, or cut in halves and remove pits. | | |
| | **To Pack Hot:** Bring to a boil in liquid and just heat through or cook about 1 to 3 minutes. Pack hot and cover with hot liquid. Seal. | 20 | 20 |
| | **To Pack Raw:** Fill the jars with uncooked apricots. Cover with boiling liquid. Seal. | 25 | 30 |
| **Berries other than strawberries** | Drain well after washing. For firm berries, add ½ cup sugar to each quart of fruit. Cover pan, bring to a boil and shake pan to keep fruit from sticking. Pack hot and cover with hot liquid. Seal. | 10 | 10 |
| | For red raspberries and other soft berries, fill jars with raw fruit and shake down for a full pack. Cover with boiling syrup made with juice or water (½ cup sugar and about ¾ cup juice for each quart). Seal. | 10 | 15 |
| **Cherries** | Wash, remove stems, sort for size and ripeness, and pit if desired. If left whole, prick to help prevent splitting. | | |
| | **To Pack Hot:** For pitted cherries, follow directions for firm berries. For cherries with pits, follow directions for firm berries but add a little water to prevent sticking. Or bring to a boil in hot syrup. Pack hot. Seal. | 15 | 15 |
| | **To Pack Raw:** Pack into hot jars. Cover with boiling syrup or juice. Seal. | 20 | 25 |
| **Figs** | Use tree-ripened figs that are not overripe. Sort and wash. Bring to a boil in hot water. Let stand in the hot water for 3 to 4 minutes. Drain. Pack hot into hot jars. Add 1 tablespoon of lemon juice to each 1-quart jar. Cover with boiling liquid. (Do not use soda in preparing figs.) Seal. | 90 | 90 |
| **Grapes** | Use ripe Muscat or slightly underripe Thompson Seedless grapes for canning. Remove stems and wash. | | |
| | **To Pack Hot:** Bring to a boil in a small amount of liquid. Pack hot into hot jars. Cover with the hot liquid. Seal. | 15 | 15 |
| | **To Pack Raw:** Put into hot jars and cover with boiling liquid. Seal. | 20 | 20 |
| **Nectarines** | Same as for freestone peaches. | | |
| **Peaches** | To peel all except canning varieties of clingstones,[3] dip in boiling water for about 1 minute, plunge into cold water, then slip off skins. Cut into halves and remove pits. To keep from darkening, dip in 1 gallon water that contains 2 tablespoons each of salt and vinegar. Drain at once. Peel canning varieties of clingstone peaches like you would apples, preferably with a stainless steel knife, or dip into a lye solution[3] until the skins loosen. When preparing a bushel or more of clingstone peaches, lye peeling saves time. Once peeled, a cut around the peach and a twisting motion between the hands will remove one-half the fruit from the pit. You can remove the pit from the second half with a special spoon-shaped knife or cut it out carefully with a paring knife. | | |

Table 5.6. Continued

| FRUIT | HOW TO PREPARE | | PROCESSING TIME (minutes) Pint | Quart |
|---|---|---|---|---|
| **Peaches (cont.)** | **To Pack Hot:** If fruit is juice, add ½ cup sugar to each quart of raw fruit. Bring to a boil. Drop less juicy fruit into a medium-thin syrup that is boiling hot. Just heat through. Pack hot. Cover with boiling liquid. Seal. | Clingstone | 20 | 25 |
| | | Freestone | 15 | 20 |
| | **To Pack Raw:** Pack in jars with the cut side down and the edges overlapping. Cover with boiling liquid. Seal. | Clingstone | 25 | 30 |
| | | Freestone | 20 | 25 |
| **Pears** | Ripen pears for canning after picking. Do not allow them to become too soft. Pears may be dipped into a lye solution to soften skins.[3] Pare, cut in halves, and trim out cores. | | | |
| | **To Pack Hot:** Same as for less juicy peaches. Seal. | | 15 | 20 |
| | **To Pack Raw:** Same as for peaches. Seal. | | 20 | 25 |
| **Plums** | Sort, remove stems, and wash. If canning whole, prick to help prevent bursting. Or cut in halves. | | | |
| | **To Pack Hot:** Bring to a boil in juice or in a thin to medium syrup. Pack hot. Cover with boiling liquid. Seal. | | 15 | 15 |
| | **To Pack Raw:** Pack the fruit into the jar. Cover with boiling juice or syrup. Seal. | | 20 | 20 |
| **Prunes, fresh** | Leave whole if small, or if large, cut in halves and remove pits. Peel if skin is tough. Prick whole fruit to help prevent checking. Cover with boiling syrup. | | | |
| | **To Pack Hot:** Same as for plums. Seal. | | 15 | 15 |
| | **To Pack Raw:** Same as for plums. Seal. | | 20 | 20 |
| **Rhubarb** | Cut into ½-inch lengths. Add ½ cup sugar to each quart of rhubarb and let stand 3 to 4 hours to draw out juice. Bring to a boil. Pack hot. Cover with hot juice. Seal. | | 10 | 10 |
| **Strawberries** | Not recommended because the produce is usually not satisfactory. | | | |
| **Tomatoes** | Sort, picking out any that are spoiled or green. (If tomatoes are excessively dirty, wash with a solution containing 4 teaspoons chlorine bleach in each gallon water.) Dip in boiling water long enough to crack skins (about 1 minute). Dip in cold water. Peel and remove cores. Save any juice to add to the tomatoes when heating. Note: if using Meyer variety lemons for juice, **add 4 teaspoons per quart** since this variety is less acid than others. | | | |
| | **To Pack Hot:** Bring whole, peeled tomatoes to a boil. Pack immediately into hot jars. Cover with the hot liquid in which the tomatoes were heated. Add 1 teaspoon salt and 2 teaspoons vinegar or 2 teaspoons lemon juice to each quart. Seal. | | 15 | 15 |
| | **To Pack Raw:** Pack raw, whole, peeled tomatoes tightly to the tops of hot jars. Press tomatoes down after each two tomatoes are added to release juice and to fill spaces. Add 1 teaspoon salt and 2 teaspoons vinegar or 2 teaspoons lemon juice to each quart. Seal. | | 30 | 30 |

[1]Adapted from York, G.K. **Home Canning of Fruits.** HXT-32, Number 2269. Cooperative Extension. University of California. Berkeley, California. 1975.

[2]Fill quart or smaller jars to within ½ inch of tops; 2-quart jars to within 1 inch of tops. Increase processing time by half for 2-quart jars.

[3]Prepare lye solution in stainless steel (not aluminum) by combining 2 gallons of **cold** water and 6 tablespoons of lye and then heating to boiling. Care is needed to avoid splattering when adding a few pieces of fruit at a time to the boiling lye solution. After 30 to 60 seconds, quickly remove fruit from lye and transfer to the sink without dripping on any surfaces. Rinse thoroughly under cold, rapidly running water to remove the skins. Can the peeled fruit at once or use dilute brine to avoid browning. Immediately dispose of lye solution by rinsing sink drain with a great deal of cold, running water or flush down the toilet, flushing several times to remove all traces from the pipes.

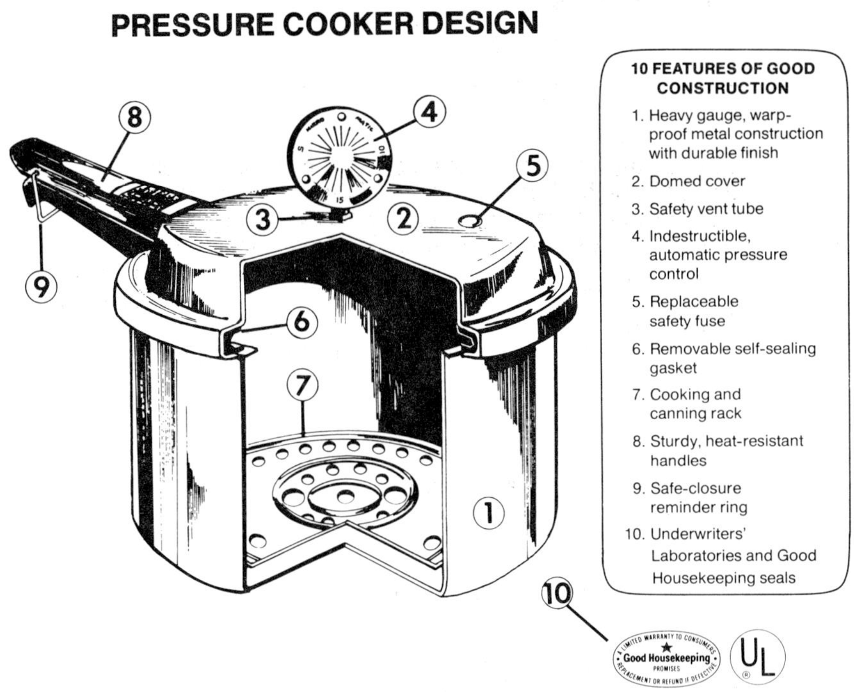


*Figure 5.23. Diagram of a pressure saucepan. This device and pressure canners are designed with the same basic engineering features.*

**Pressure canner**

This piece of equipment offers the only device in the home which can assure that the final temperature of 240°F is reached. Careful attention needs to be given to following the specific directions outlined by the manufacturer for the specific canner being used.

Ordinarily, preparation for pressure canning includes putting at least two inches of boiling water and not more than three inches in the bottom of the canner. The filled jars then are placed on the rack in the canner, being careful to maintain at least a little space between each jar. This spacing is essential for permitting the steam to circulate freely to all areas of the canner during processing. As soon as the jars are all in place, the lid is fastened securely in its locked position. However, the petcock is left open while heat is applied to the canner and is allowed to remain open until the canner has been exhausted, i.e., steam is coming continuously from the open petcock.

It is very important to exhaust the pressure canner. If this step is not done carefully and completely, the pressure indicated by the gauge will not accurately indicate the actual situation within the canner, and the jars will not be processed at a high enough temperature. This situation obviously presents a risk since the foods being canned in a pressure canner (vegetables, meats, poultry, and fish) may contain heat-resistant spores which can produce toxins during storage if the spores are not killed during processing.

Exhausting is done by heating the unpressurized pressure canner until only active steam remains in the space above the water. First the canner is heated until a steady stream of steam issues from the canner. Heating is then timed for ten minutes to insure complete exhaustion. At the end of this heating period, the petcock is closed. The pressure is allowed to reach ten pounds at sea level before timing of processing is begun. As soon as the gauge indicates that the desired pressure has been reached, timing of processing is begun. The pressure needs to be maintained accurately at ten pounds throughout the processing period until the correct time has elapsed. Immediately the canner is removed from the heat and allowed to cool to relieve the pressure without any disturbance.

*Figure 5.24. Approximately two inches of water are placed in a pressure canner in preparation for the addition of the jars of food to be processed.*

*Figure 5.25. The cover of the pressure canner is locked tightly in preparation for building the necessary pressure to achieve the desired temperature of 240°F. in the canner during processing.*

*Figure 5.26. The pressure canner is exhausted by heating the pressure canner until steam pours from the petcock in a continuous stream for 10 minutes (or according to manufacturer's directions).*

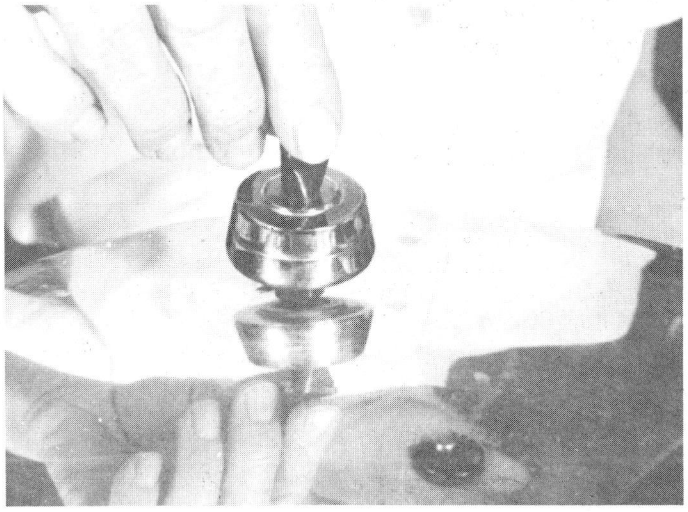

*Figure 5.27. After steam has issued continuously from the petcock for ten minutes, the petcock is closed, and pressure begins to build in the canner.*

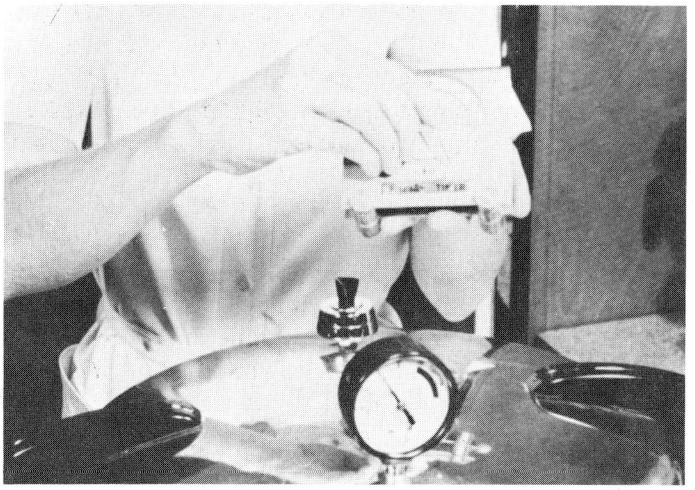

*Figure 5.28. When the pressure gauge indicates that ten pounds of pressure have built up in the canner, the timer is set for the processing time required for the specific food being canned. The heat is regulated under the canner to maintain exactly ten pounds of pressure.*

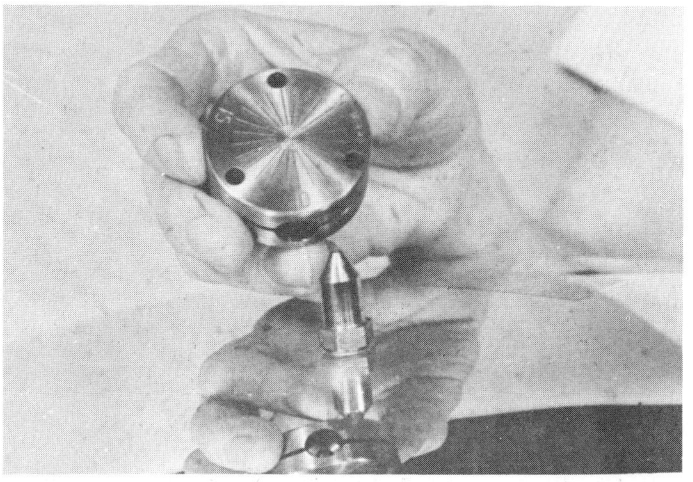

*Figure 5.29. If the pressure canner has a weighted gauge, the gauge is positioned after the canner has been exhausted, as described in Figure 5.26. Care must be taken to be sure that the gauge is placed for the desired amount of pressure (usually ten pounds).*

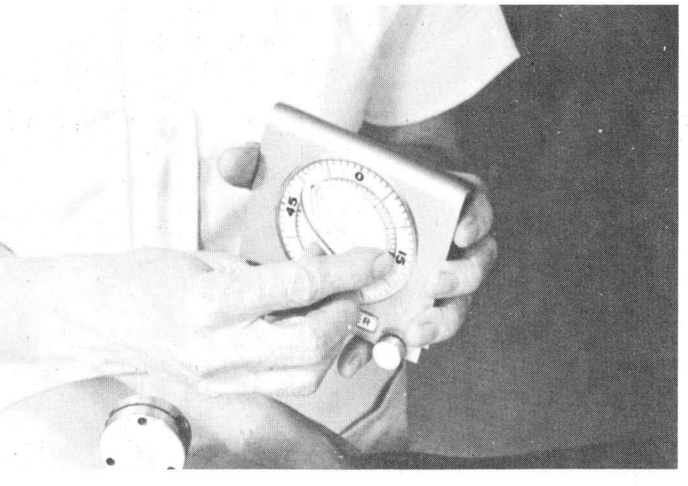

*Figure 5.30. The pressure is built up in the canner until the gauge jiggles at the rate of about three times per minute. Then the timer is set for the processing time required. Attention must be given to maintaining the jiggle of the gauge at the rate of approximately three times per minute. The heat is adjusted below the canner to regulate the rate.*

The processing times recommended for vegetables are outlined in Table 5.7. The adjustments for processing at altitudes are described in Table 5.8.

At elevations above sea level it is necessary to make adjustments in the amount of pressure used for canning low-acid foods. This modification is needed in order to reach the desired temperature of 240°F. If the pressure canner has a pressure gauge, the pressure should be adjusted upward, according to the data presented in Table 5.8. The increase in pressure will bring the temperature in the pressure canner to the desired 240°F. When a weighted gauge is used, it is necessary to either (1) have the weighted gauge corrected for the altitude by the manufacturer, or (2) do any processing above 2,000 feet elevation at 15 pounds of pressure to insure an adequate temperature for safe processing.

**Pressure saucepan**

A pressure saucepan can be used if only small amounts of food need to be canned. The capacity is too limited to make this utensil suitable for canning low-acid foods in substantial quantities. The pressure saucepan is an excellent piece of equipment to use when canning for the first time. This enables people to try canning before investing in a large canner. Jars with a capacity of half a pint fit well into a pressure saucepan. The manufacturer's directions for the saucepan should be observed when this device is used.

Table 5.7. Guide to pressure canning of vegetables [a,b]

| VEGETABLES | HOW TO PREPARE | PROCESSING TIME (minutes) Pint | Quart |
|---|---|---|---|
| **Artichokes** | Use small artichokes. Trim to 1¼ to 2 inches in length. Pre-cook 5 minutes in water to which ¾ cup of vinegar per gallon has been added. Drain. Pack hot into hot jars. Do not overfill. Cover with a boiling brine prepared by adding ¾ cup of vinegar or lemon juice and 3 tablespoons salt to 1 gallon water. Fill to within ½ inch of tops of pint or quart jars. Seal. | 25 | 25 |
| **Asparagus** | Sort, wash, and cut in lengths ¾ inch shorter than jar. Cut off scales (bracts). Use ½ teaspoon salt for pint jars; 1 teaspoon salt for quart jars. | | |
| | **To Pack Hot:** Pre-cook 1 to 3 minutes in boiling water to wilt. Then plunge quickly into cold water. Pack at once. Seal. | 28 | 32 |
| | **To Pack Whole:** Gather a bundle of stalks with cut ends down and fill jar. Do not pack tightly. Seal. | 28 | 32 |
| | **To Pack Cuts:** Fill jar. Add salt. Add boiling water to within ½ inch of jar tops. Seal. | 28 | 32 |
| **Beans, fresh lima** | **To Pack Hot:** Proceed as for peas. Process as directed for baby limas. **To Pack Raw:** Use ½ teaspoon salt for pint jars: 1 teaspoon salt for quart jars. **Baby limas** — pack pint jars loosely to within 1 inch of tops; quarts to 1¼ inches. **Large beans** — pack pint jars loosely to within ¾ inch of tops; quarts to 1¼ inches. Add salt. Fill to top with boiling water. Seal. | 40 | 50 |
| **Beans, string** | Sort and snip or string if necessary. Use 1 teaspoon salt for pint jars; 1 teaspoon salt for quart jars. For mature beans, see **peas, mature.** | | |
| | **To Pack Hot:** Cut in 1- to 1½-inch lengths. Pre-cook in boiling water until pliable, about 2 to 5 minutes. Pack hot into hot jars. Add salt. Cover to within ½ inch of jar tops with the boiling liquid in which the beans were pre-cooked. Add boiling water if needed. Seal. | 20 | 25 |
| | If beans are left whole, pack beans standing on ends. Seal. | 25 | 30 |
| | **To Pack Raw:** Cut into 1-inch pieces. Pack tightly to within ½ inch of jar tops. Add salt. Cover with boiling water to within ½ inch of top. Seal. | 20 | 25 |
| **Beets** | **To Pack Hot:** Leave on roots and 1 to 1½ inch of stems. Parboil until skins slip off (about 15 minutes). Dip in cold water. Peel, trim, and slice. Discard woody beets. Re-heat in small amount of water. Pack hot into hot jars. Add ½ teaspoon salt to pint jars; 1 teaspoon salt to quart jars. Cover to within ½ inch of jar tops with the boiling liquid in which the beets were re-heated. Add boiling water if needed. Seal. | 35 | 40 |
| | **Raw packing and pickling of beets is not recommended.** | | |
| **Carrots** | **To Pack Raw:** Wash and scrape or peel. Pack cold, sliced, or asparagus style to within 1 inch of tops of pint or quart jars. Add ½ teaspoon salt to pint jars; 1 teaspoon salt to quart jars. Fill to top with boiling water. Seal. | 30 | 30 |
| **Celery** | Prepare and slice. Use ½ teaspoon salt for pint jars; 1 teaspoon salt for quart jars. | | |

Table 5.7. Continued

| | | | |
|---|---|---|---|
| **Celery (cont.)** | | | |
| | **To Pack Hot:** Pre-cook 1 to 3 minutes, depending on size and tenderness. Pack hot into hot jars. Add salt. Cover to within ½ inch of jar tops with boiling liquid in which the celery was pre-cooked. Add boiling water if needed. Seal. | 35 | 35 |
| | **To Pack Raw:** Slice or cut asparagus style. Pack loosely to within ½ inch of jar tops. Add salt. Cover with boiling water to within ½ inch of tops. Seal. | 30 | 30 |
| **Corn whole kernel** | **To Pack Hot:** Can very soon after harvest. Use a sharp knife to cut raw corn from cob to two-thirds of the total depth of the kernels. Do not scrape the cobs. Cover well with brine (1 level tablespoon salt to 1 quart water). Heat to boiling point. Pack hot into jars to within 1 inch of jar tops. Seal. **Raw pack is not recommended.** | 55 | 70 |
| **Corn, cream style** | **To Pack Hot:** Prepare as for whole kernel corn, but scrape the cobs (do not scrape off any of the cob material). Proceed as directed for whole kernel corn. Leave 1¼ inches headspace. Quart jars are not recommended. Seal. **Raw pack is not recommended.** | 85 | – |
| **Greens: spinach, swiss chard, beet greens, other greens** | Home canning is not recommended. | | |
| **Onions, small white** | Follow directions for artichokes. | | |
| **Peas** | Use ½ teaspoon salt for pint jars; 1 teaspoon salt for quart jars. | | |
| | **To Pack Hot:** Can only young, tender peas. Hull and pre-cook for 1 to 4 minutes in a small amount of water until the skins wrinkle. Pack hot into hot jars to within 1¼ inches of tops. Add salt. Cover to within 1 inch of jar tops with the boiling liquid in which the peas were cooked. Add boiling water if needed. Seal. | 40 | 45 |
| | **To Pack Raw:** Pack loosely to within 1 inch of jar tops. Add salt. Cover with boiling water to within 1 inch of top. Seal. | 40 | 45 |
| **Peas, mature** | **To Pack Hot:** Follow directions for peas. Fill the jars less full of peas and add liquid to within 1 inch of jar tops. Seal. **Raw pack is not recommended.** | 50 | 55 |
| **Peppers, bell (green and red) and pimiento** | **To Pack Hot:** Cut out the stem end of each pepper and remove the core and seeds. Peel peppers by heating in a gas flame or roasting in a very hot oven until the skins separate. Chill at once in cold water. Pack into jars. Cover with boiling water to within ½ inch of jar tops. Add ½ teaspoon salt to pint jars; 1 teaspoon salt to quart jars. It is also necessary to add 1 tablespoon lemon juice to each quart jar. Process at only 5 pounds pressure: higher pressures affect texture and flavor. Seal. **Raw pack is not recommended.** | 50 | 60 |
| **Potatoes, new** | Scrape the peel off new potatoes. Leave small ones whole; cut larger ones in halves. Pack cold without pre-cooking. Add boiling brine made with 1½ to 2 tablespoons salt to 1 quart water. Fill to within ½ inch of jar tops. Seal. | 35 | 40 |
| **Pumpkin or mature squash** | Scrape out fibrous material and cut flesh and rind into strips. Boil in water, or steam, until flesh is soft. Scrape flesh from rind and press through a colander. Bring to a boil. Pack hot into jars to within ½ inch of tops. Add ½ teaspoon salt to pint jars; 1 teaspoon salt to quart jars. Seal. **Raw pack is not recommended.** | 85 | 115 |
| **Squash, banana** | Peel and steam or cook in a small amount of water. Mash or press through a colander and proceed as directed for pumpkin. Seal. **Raw pack is not recommended.** | 85 | 115 |
| **Turnips** | Follow directions for carrots. | | |

[a] Adapted from York, G.K. Home Canning of Vegetables. HXT-54, Number 2270. Cooperative Extension . University of California. Berkeley, California. 1975.

[b] Process at 10 pounds pressure (240°F.). If the pressure drops below 10 pounds for longer than 5 minutes, time the cooking over again. If the pressure drop is for less than 5 minutes, process for an additional 10 minutes.

Table 5.8. Pressure adjustments for altitude

| Altitude (feet) | Pressure (pounds) |
|---|---|
| 2,000 | 11 |
| 4,000 | 12 |
| 6,000 | 13 |
| 8,000 | 14 |
| 10,000 | 15 |

The time required for processing meats, poultry, and fish (Table 5.9) is substantially longer than that required for vegetables. Processing is done at ten pounds of pressure, with necessary adjustments for altitude corresponding to those made for pressure canning vegetables (see Table 5.8). Various recipes for meats in combinations also can be canned. The processing times needed for these will vary with the types of ingredients included. When a thick sauce, such as one thickened with flour, is included, the processing time will need to be lengthened because viscous or highly bulky items do not transmit heat through the jar nearly as efficiently as more fluid substances do.

Table 5.9. Guide for canning meats, poultry, and fish at 240°F

| Item | Processing time in pressure canner (in minutes) Pints | Quarts |
|---|---|---|
| Lamb | 75 | 90 |
| Veal | 75 | 90 |
| Beef | 75 | 90 |
| Pork | 75 | 90 |
| Chicken | 75 | 90 |
| Turkey | 75 | 90 |
| Fish[1] | 110 | |
| Crab | 65 | |

[1]Fish should not be canned in jars larger than a pint because of the extremely long processing time required.

## Microwave Oven Canning

Canning in the microwave oven is not recommended at the present stage of technology. The pattern of microwave energy in the oven is not uniform throughout the cavity in today's models. Furthermore, pressure can be built to such a high level in the jars that they will explode if the heating time is too long. Because of these hazards, this appliance should not be used for canning any foods.

## Oven Canning

In some foreign publications as well as in some older publications in this country, the suggestion may be encountered that food should be canned by processing in the oven. This practice definitely should be avoided. There is the possibility of explosion of jars in oven canning. Furthermore, the contents of the jar are not heated to a high enough temperature to assure safety of the food.

# TREATMENT FOLLOWING PROCESSING

## Cooling the Product

### Fruits

As soon as processing of boiling-water-bath canned foods has been completed, the jars are removed immediately from the canner with the aid of a jar lifter. If this device is not available, remove enough water from the bath to expose the upper portion of the jars enough so that they can be removed safely using a pot holder. A ladle or cup works well for removing the water without unnecessary risk. The jars should be transferred to a rack or folded cloth, leaving ample space between jars so that air currents will help to cool them efficiently. They should not be placed on cold surfaces or exposed to cold drafts because of the thermal shock these situations create on the very hot glass of the jars. If as much as an inch and a half of the liquid has been lost from the jar during processing, the food should be refrigerated and used within a couple of days.

The treatment of the lids following processing varies with the type of closure used. Most closures should remain undisturbed until the product has cooled completely so that the seal effected by the sealing compound will not be broken. However, the porcelain-lined zinc lids need to be tightened just as soon as the jars are removed from the water bath. The manufacturer's directions for the closures should be followed in all instances.

### Vegetables and meats

Foods processed in a pressure canner must remain in the pressure canner while the vessel cools gradually, and the pressure gauge drops to zero. After the gauge indicates zero, five minutes should elapse before the petcock is opened or the weighted gauge is removed. Then wait another 15 minutes after the petcock has been opened before opening the lid. The lid should be lifted so that the steam will be directed away from the operator. The jars are allowed to remain in the uncovered canner for another 10 to 15 minutes before being transferred to a rack or cloth, as described for fruits above. The same precautions regarding loss of liquid and the tightening of zinc lids need to be observed, also.

## Testing for Sealing

When the jars have cooled, the metal bands can be removed from the two-piece metal lids. Some variations of this type of closure may need to have

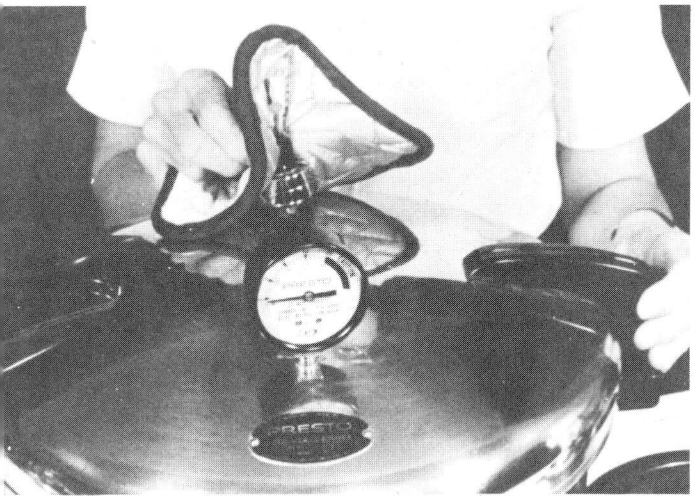

*Figure 5.31. After the processing time has elapsed, the canner is allowed to cool without any attempt to hasten the cooling process. When the pressure has dropped to atmospheric pressure in the canner, the petcock is opened.*

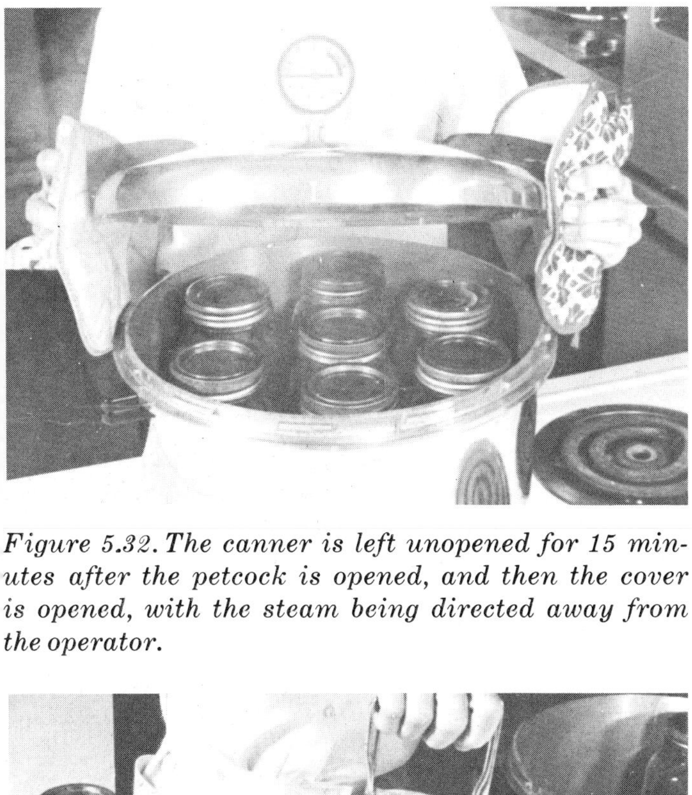

*Figure 5.32. The canner is left unopened for 15 minutes after the petcock is opened, and then the cover is opened, with the steam being directed away from the operator.*

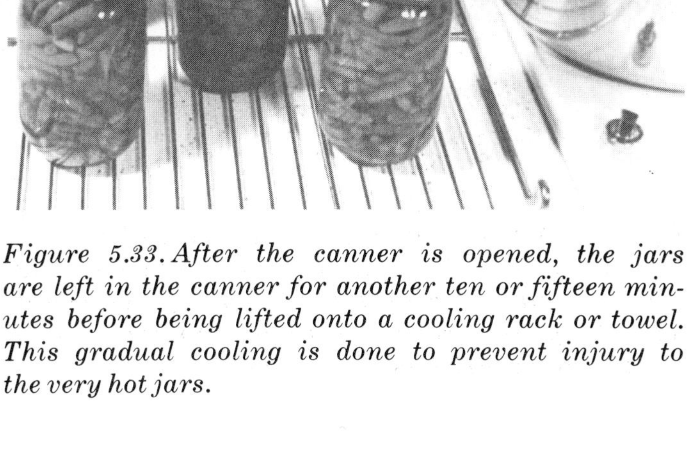

*Figure 5.33. After the canner is opened, the jars are left in the canner for another ten or fifteen minutes before being lifted onto a cooling rack or towel. This gradual cooling is done to prevent injury to the very hot jars.*

*Figure 5.34. When the jars have cooled, the ring of this type of two-piece closure is removed in preparation for storing.*

*Figure 5.35. If a jar has sealed properly, the lid will be drawn downward in the center.*

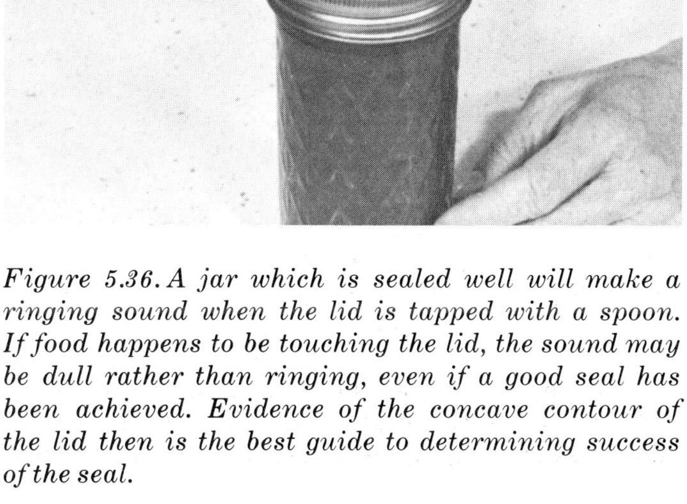

*Figure 5.36. A jar which is sealed well will make a ringing sound when the lid is tapped with a spoon. If food happens to be touching the lid, the sound may be dull rather than ringing, even if a good seal has been achieved. Evidence of the concave contour of the lid then is the best guide to determining success of the seal.*

the band retained on the jar. Directions should be followed.

Each jar should be checked to be sure that it has sealed properly. The lid of two-part metal covers should be slightly concave as a result of the small vacuum created in the jar when it has been processed and sealed properly. An additional check is provided by tapping the top with a spoon. If the sound seems to be rather clear, rather than dull, the jar probably is sealed well. Yet another test is to press down on the lid with the finger. A sealed lid will not yield.

The remaining step prior to storage is simply to label the jar, being sure to include the date (and year).

*Figure 5.37. Canned goods should be labeled with the date and any pertinent description which will help in identifying the product following storage.*

### Reprocessing

It is imperative that a good seal be achieved before the canned food is stored. If any jars do not have a good seal with the closure, they should either be reprocessed immediately or placed in the refrigerator for early use. Reprocessing is done by following the procedures outlined for the specific food being canned. The timing and all steps are the same as they would be if the food had not been processed before.

### Storing the Product

Canned foods ideally are stored in a cool, dark, dry place until they are used. This circumstance helps to promote maximum shelf life. Storage at a temperature of 18°C (75°F) cuts shelf life to half that of canned goods stored at 60°F. Under good storage conditions, properly processed canned foods should have a safe shelf life of two years; highly acidic foods may begin to deteriorate in quality after a year and a half of storage. Although canned foods can be stored these extended lengths of time, generally it is a good practice to plan to use canned foods within a year. This plan facilitates inventory maintenance and assures rotation of foods by freeing storage space for the new crop annually.

*Figure 5.38. All low-acid foods (vegetables, fish, poultry, and meat) should be placed directly from the jar into a saucepan without anyone even touching or tasting the contents.*

### Using the Canned Food

Home canned foods classified as high-acid foods (fruits) can be used directly from the jar, just as commercially canned fruits would be used. Low-acid foods present a possibly lethal hazard if heat processing has not been adequate. As a safety measure, all low-acid foods (vegetables, meat, and poultry) should be poured directly into a saucepan and boiled for 15 minutes before even being tasted. No low-acid food should be tasted before this boiling period has elapsed.

*Figure 5.39. Low-acid foods should be boiled vigorously for 15 minutes as a safety precaution to avoid any possibility of botulism. The boiling period will render any toxin which might have been produced by* Cl. botulinum *harmless.*

## NUTRITIVE VALUE OF HOME CANNED FOODS

Some loss of nutrients does occur during canning of foods. Some loss of ascorbic acid and the B vitamins is noted, largely as a result of losses in water blanching and during heat processing. However, these foods are highly reliable sources of nutrients, particularly if very fresh produce has been used in canning.

## EVALUATION OF CANNED FOODS

Canned foods should be attractive in the jar and when served. The texture should be soft, yet firm. The color should be bright and appealing, and the flavor should be fresh. Canned chlorophyll-containing vegetables will always be an olive-green color because of the prolonged heating period required for safety.

Before opening jars of food, the container should be checked for any evidence of spoilage. The lid of the two-piece closures should be depressed slightly, indicating a good seal still exists. There also should be no evidence of leakage around the seal. Bulging of the jar lid is a clear indication of gas production within the jar, which is caused by improper processing (either inadequate heat treatment or failure to seal). This may be observed in low-acid foods.

### Flat-Sour Spoilage

The type of spoilage known as flat-sour is caused when viable acid-producing microorganisms remain in the canned food. *Bacillus coagulans* is the microorganism commonly responsible for spoilage of this type. This anaerobic bacillus converts sugar to lactic acid without producing any gas. The result is a more acidic food product, but one without gas being produced to signal the problem.

### Sulfide Spoilage

Foods contamined with viable *Clostridium nigrificans* will contain hydrogen sulfide, a gas which can be detected readily by its odor when the jar is opened. Sometimes a slightly bulging lid will signal presence of this problem, but this is not always the case.

### Hydrogen Swells

If there is no evidence of spoilage other than a bulging container, the problem probably is caused by hydrogen gas being generated as the result of an acid food interacting with the metal of the container. This condition, known as "hydrogen swells", is only found in home canned foods in metal cans. The generation of hydrogen gas is caused by the presence of thermophilic anaerobes. Sometimes an odor of butyric acid may accompany the spoiled product.

### Botulism

Spoilage due to production of a paralyzing exotoxin by spore-forming *Clostridium botulinum* cannot be noted by examination. Because of its high degree of toxicity, the low-acid foods which could contain the toxin are never tasted until the canned food has been boiled for 15 minutes. The types of botulism most commonly diagnosed are: Type A, Type B, and Type E. The source of Types A and B is usually from dirt, whereas Type E is basically either from marine or inland lake and stream waters. Only a few cases of botulism have been attributed to Type C and to Type F (apparently introduced in the food via water). The spores of Types A and B are the most resistant to heat processing, while Type E spores are less resistant. Type C spores are intermediate in resistance between Type E (and also Type F, which is similar to Type E), and the very resistant Types A and B. Type E is of greatest concern in canned fish because of its occurrence in aquatic food sources.

## SUMMARY

Fruits, vegetables, meats, poultry, and fish can be canned satisfactorily in the home providing attention is given to using correct processing techniques and equipment. The decision on the type of processing required for food safety is based on the pH of the food. Foods with a pH lower than 4.5 can be processed safely in the water bath canner under atmospheric conditions. Fruits, pickles, and jams can be processed this way, following preparation using either a hot or cold pack, usually with a sugar syrup being added. Foods with a pH above 4.5 include vegetables (with the exception of tomatoes), meats, poultry, and fish. These foods need to be processed at 10 pounds pressure (or more at higher elevations) so that they are heated in an environment of 240°F. Careful timing of heat processing is required regardless of the method of canning. Special precautions for cooling the pressure canner and its contents slowly must be observed to avoid extreme and sudden pressure changes in the jar and its surroundings. Microwave ovens and convention-

*Figure 5.40. Fruits are relatively acidic and are well suited to canning by water bath canning at atmospheric pressure.*

*Figure 5.41. Vegetables and meats require pressure canning in order to reach the high temperature necessary to be certain that the spores of* Cl. botulinum *have been killed during processing.*

al ovens should not be used for canning because of safety hazards.

Canned foods should be assessed visually before opening to note any signs of spoilage or leakage. In addition, low-acid foods should be boiled 15 minutes before they even are tasted because of the possible presence of the toxin from *Cl. botulinum.*

## BIBLIOGRAPHY

Ball Corporation. *Ball Blue Book.* Ball Corp. Muncie, Indiana. 29th ed. 1972.

Cameron, E.J. and J.R. Esty. Comments on the microbiology of spoilage in canned foods. *Food Res. 5*: 549. 1940.

Consumer and Food Economics Research Division. *Home Canning of Fruits and Vegetables.* Home and Garden Bulletin No. 8. U.S. Dept. of Agriculture. Washington, D.C. 1971.

Goldblith, S.A., et al. *Introduction to Thermal Processing of Foods.* Avi Publishing Co. Westport, Conn. 1961.

Groppe, C.C. and R. Crawford. *Canning and Freezing Fish at Home.* Cooperative Extension. University of California. Berkeley, California. 1974.

Harris, H. and L.M. Davis. *Use of Low Water Level in Boiling Water Bath Canning.* Agricultural Experiment Station. Auburn University. Auburn, Alabama. 1976.

Joslyn, M.A. *Bottling, Canning, Freezing Fruit Juices and Tomato Juice.* Leaflet 4002. Cooperative Extension. University of California. Berkeley, California. 1974.

Kerr Glass Manufacturing Co. *Kerr Home Canning and Freezing Book.* Sand Springs, Oklahoma. 1974.

Kuhn, G.D. and L.W. Hamilton. *Sealing Performance of Jar Lids for Home Canning.* Agricultural Experiment Station, University Park, Pennsylvania. 1976.

Stumbo, C.R. *Thermobacteriology in Food Processing.* Academic Press. New York. 2nd ed. 1973.

York, G.K. *Home Canning of Fruits.* HXT-32, Number 2269. Cooperative Extension. University of California. Berkeley, California. 1975.

York, G.K. *Home Canning of Vegetables.* HXT-54, Number 2270. Cooperative Extension. University of California. Berkeley, California. 1975.

- Tomatoes are fruits - cuz of high acid content -
add tsp. of salt for flavor - not preservative -
pts. 35 min
qts 45 min -

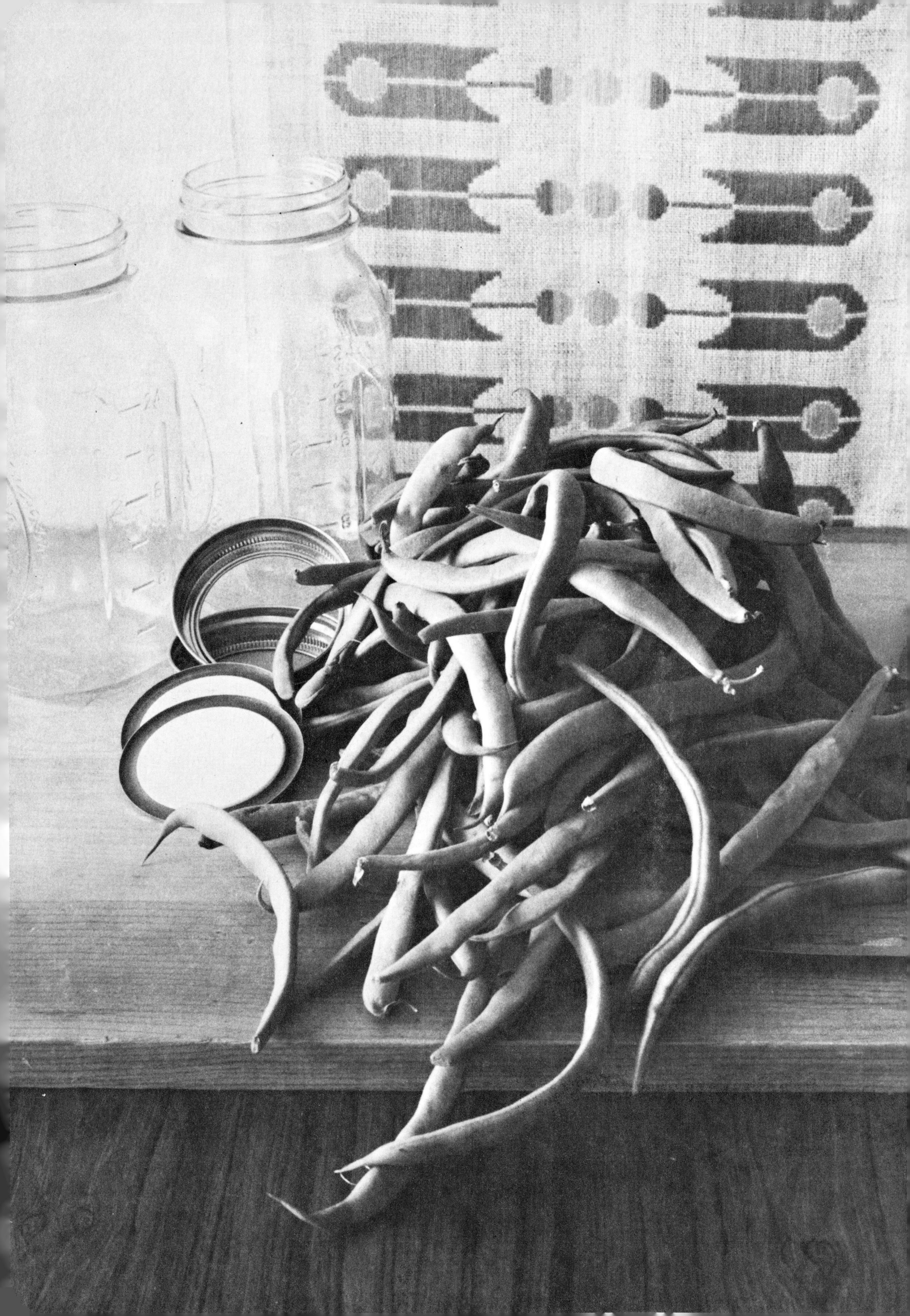

# Chapter 6. Recipes for Canning

## PRELIMINARIES

### Pressure Saucepans and Canners

Canning of vegetables begins with selection of equipment which will provide a reasonable capacity for canning jars under pressure. The smallest device available for consideration is a 4 quart pressure saucepan, which needs to be equipped with a pressure gauge calibrated for maintaining 5 and 10 pounds of pressure. It is possible to can half-pint jars in a 2½-quart pressure saucepan, but this size is too small for convenience in canning most foods. In contrast, the 8-quart capacity pressure canner is deep enough to permit the processing of quart jars. However, persons with large quantities of food to can will find it far more convenient to use pressure canners with larger capacities. These are practical for individuals who regularly can large quantities of foods annually. The canners are large and require a considerable amount of storage space and, thus, may not be a wise choice for persons with limited storage space or with minimal amounts of food to can. The capacities of appropriate cookers and canners are presented in Table 6.1.

*Figure 6.1. A pressure saucepan can be used satisfactorily for processing small jars of food and is a convenient means of studying basic techniques in home canning.*

Table 6.I. Canning capacities of selected pressure canners

| Cooker size | ½-pint jars (0.24 l.) | Pint jars (0.47 l.) | Quart jars (0.95 l.) |
|---|---|---|---|
| 2½ qt. (2.3 l.) | 5 | – | – |
| 4 qt. (3.8 l.) | 5 | 4 | – |
| 6 qt. (5.7 l.) | 8 | 7 | – |
| 8 qt. (7.6 l.) | 8 | 7 | 4 |
| 12 qt. (11.4 l.) | 13 | 10 | 7 |
| 16 qt. (15.2 l.) | 26 | 10 | 7 |
| 22 qt. (20.8 l.) | 38 | 20 | 7 |

### Design and Maintenance Features

Certain features of a pressure saucepan or pressure canner should be checked before using these devices for canning. The rubber gasket is essential for assuring a good seal between the canner and its cover. This gasket must always be clean and in good condition, that is, it must fit well rather than being shrunken or stretched. This gasket is replaceable, and a new replacement should be obtained if the original one becomes misshapen. Also, it is important to be certain that the handles are fastened securely to the base and cover.

The pressure control device also must be in excellent working order. If the canner has a one-piece weighted pressure device, it is important to be sure that there are no food particles plugging it. No calibration is required on this type of pressure device, but an annual check should be made on the accuracy of the dial-type gauges found on some pressure canners. The dial gauges can get out of calibration, an error which will cause canning to be done possibly at too high a temperature or at too low a temperature. The first instance causes unnecessary overcooking of food, a circumstance leading to less appealing finished products with overly soft textures. The latter condition, that of too low a temperature in the canner, presents the possibility of unsafe canned foods because of failure to reach the high temperatures needed for destroying harmful microorganisms in canned vegetables. Dial pressure gauges can be checked at a limited number of service centers. Information on the local center can best be obtained by contacting the County Extension Service.

Pressure saucepans and canners are equipped with a safety valve or safety fuse. These can and should be replaced if they pop out. Rubber safety valves should be replaced at the same time the gasket is replaced because the gasket serves as a good indicator of the deterioration of the rubber in the safety fuse as well as in the gasket.

Some safety devices consist of a rubber valve with a metal plunger in the center. This design feature functions by venting the air in the pressure canner through the valve at first. Then the plunger rises to seal the opening, and pressure will begin to build within the canner. The safety valve, whether featuring this design or simply a single metal or rubber plug, is designed to remain in position up to a pressure of 25 pounds. If pressure begins to exceed this maximum, the safety fuse will pop out and relieve the pressure. It is apparent from this description of its function that the safety fuse must be in functional condition to provide the safety which it has been designed to insure.

*Figure 6.2b. Other pressure canners have dial gauges to indicate pressure. Dial gauges should be checked annually for accuracy.*

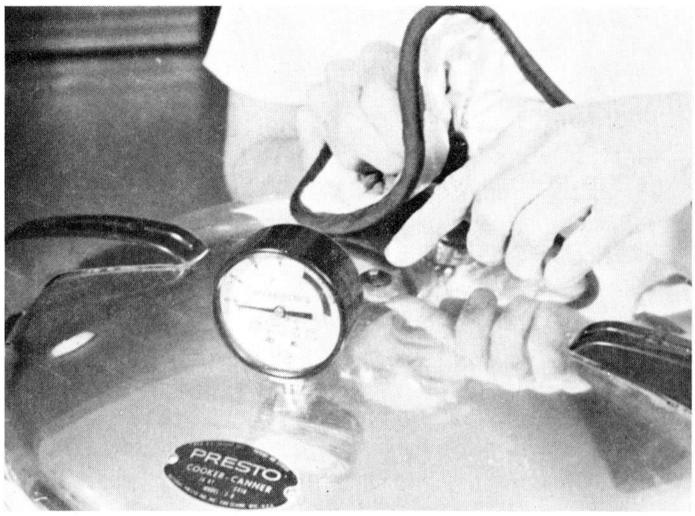


*Figure 6.3. A safety valve is a feature of the design of pressure canners. This very small device is an essential safety mechanism and must be functional.*

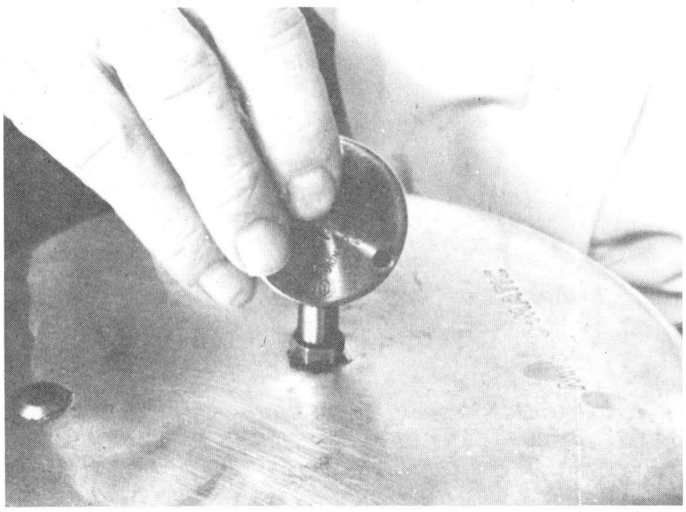

*Figure 6.2a. Some types of pressure canners are equipped with weighted pressure gauges.*

*Figure 6.4. The small vent in the petcock can become plugged with food particles; a skewer can be inserted in it quickly each time the canner is used to be certain that it is open.*

## Operation of Pressure Canners

Operation of pressure saucepans and canners can be done safely and efficiently if a few basic guidelines are kept in mind and always followed. These guidelines are as follows:

1. Place the rack in the bottom of the cooker.
2. Pour in a minimum of an inch of water in the bottom of a small pressure saucepan and a minimum of two inches of water in a large pressure canner. If desired, a tablespoon of vinegar can be added to avoid discoloration of aluminum canners.
3. Place the properly filled canning jars on the rack. The sides of the jars should not touch. If a second layer is to be stacked in a large canner, these jars must be placed on a second rack and their position should be staggered in relation to the position of the lower tier of jars.
4. Be certain that the vent opening, the rim of the canner, and the rim of the lid are perfectly clean. Position the cover on the cooker (with the rubber gasket positioned properly) and tighten according to the manufacturer's directions.
5. Pressurize the canner according to manufacturer's directions. Pressurizing requires that first the contents of the canner be heated and then that steam pressure be generated. In some canners, the vent will be left open until steam flows continuously from the vent for 10 minutes, at which time the petcock is closed or the weighted gauge is put in position. Some canners are designed to have the weighted gauge put into position as soon as the cooker is closed. With either type of design, a high heat is used to pressurize the canner as rapidly as possible. The length of time required for the container to be pressurized will be influenced by the size of the canner, the number of jars being processed, the size of the jars, and the temperature of the food and the water when heating begins. A four-quart saucepan and its contents may reach processing pressure in 10 minutes, while a 22-quart pressure canner may require an hour to become pressurized if it is filled to capacity.
6. Actual processing time is counted only when processing pressure has been reached. For equipment featuring a dial gauge, this means keeping a watchful eye on the gauge so that timing will be initiated as soon as the desired pressure is indicated. Weighted gauges indicate the correct pressure when they begin to jiggle. To avoid confusion on timing, it is best to set a timer when pressure is reached. However careful timing can be done using a clock.
7. During the processing period, the pressure in the pressure cooker is maintained at the proper level by controlling the heat input. The heat source should be adjusted downward so that the correct pressure is indicated on the dial gauge or so that the weighted pressure control jiggles between one and four times per minute. It may be necessary to make minor adjustments in the heat input during the course of processing to maintain these pressure indicators.
8. As soon as the correct processing time has elapsed, the heat should be turned off. If an electric range is being used, the cooker needs to be removed from the heating element to prevent continuing residual heat input from the cooling element. The pressurized canner is cooled at room temperature, with no attempt being made to cool it or to depressurize it abnormally rapidly. If cold water is run on the outside of the canner, or if the pressure control is removed prematurely, liquid may be forced from the jars, or the jars might even crack due to thermal shock. Approximately 25 minutes will be required to cool small pressure saucepans and about 45 minutes will be sufficient for cooling the large containers.
9. When the suggested cooling time has elapsed, the pressure control is pushed gently to determine whether the canner has reached normal pressure. If steam spurts out, the canner needs to cool longer. If no steam is evident, but a hissing sound is heard, it is safe to remove the pressure control. The hissing is simply air rushing through the vent into the chamber.
10. When the pressure control has been removed, the canner can be opened. This is done by rotating the handles to the open position, as indicated by the manufacturer. The lid is opened away from the operator so that the escaping steam will not come in contact with the skin and cause burns.
11. The jars are lifted from the opened canner with the aid of a jar lifter and are placed on a cooling rack with enough space between the jars to permit good air circulation. If a cooling rack is not available, the jars can be placed on a few layers of towels. The important point is to be careful to avoid placing jars directly on the counter or in cold drafts. Either of these conditions can place too much stress on the very hot jars.
12. As soon as jars are removed from the canner, lids which require tightening should be screwed slowly, but firmly into position. The bands on self-sealing lids are *not* tightened. If liquid has been lost from the jar during processing, no attempt is made to replace the lost liquid. If additional liquid were to be added, it would be necessary to do the entire processing step again.
13. The screw bands from self-sealing lids are removed after the processed foods have been allowed to sit for at least 12 hours. However, other lid designs do not require any additional steps at this time.
14. The seal of each jar needs to be tested. One-piece lids should be low in the center. Two-

piece lids can be pressed to be sure that they are curved down and that they do not move. A further check can be made by tapping the center of the cover with a spoon. A clear sound is an indication that a seal has been achieved.

## VEGETABLES

As described in the preceding chapter, it is necessary to employ pressure canning methods for canning low-acid vegetables so that the spores which might be present from *Clostridium botulinum* will not be viable in the canned product on the shelf. For all vegetables except tomatoes this hazard dictates the need for pressure canning to reach a safe processing temperature. Careful attention to all aspects of sanitation, to pressure control, and to accurate timing are essential to producing canned products which are of optimal quality and safe to eat.

The accompanying sequence of photographs illustrates the sequence of steps necessary for safe processing of vegetables. Green beans are used as the example, but the same basic procedures are employed for processing any vegetables other than tomatoes.

### Green Beans

#### Preparation for canning

Preparations for canning green beans, like for all other canning, begin with the assembling of the necessary equipment and supplies. In the case of green beans (and also other low-acid vegetables), a pressure canner will be required. A jar lifter also will be a valuable piece of equipment for protecting against burns when lifting the hot jars.

The jars and closures need to be examined carefully. Particular attention should be given to the condition of the rim of the jar. It is imperative that there not be any imperfections in the rim; nicks make it impossible to achieve the seal needed to keep food safe during storage following canning. Closures also must be in good condition. If zinc lids and rubber rings are being used, the rubber rings need to be in good condition, with no signs of brittleness or deterioration. The porcelain lining should be unchipped, and the zinc lid should be undamaged. If reusable two-piece closures are utilized, it is imperative that both parts be in good condition and that the sealing portion be sufficiently resilient to effect a good seal. The standard two-piece closures need to have new lids each time because the sealing compound on the gasket is only capable of sealing one time. However, the rings can be reused unless they are bent or rusted. All closures (even new ones) need to be examined for adequacy of sealing compound and any other possible manufacturing imperfections.

The jars and closures are washed very thoroughly in hot, soapy water and then rinsed well under

*Figure 6.5. A funnel, pie plate, ladle, and jar lifter are convenient items to aid in filling canning jars.*

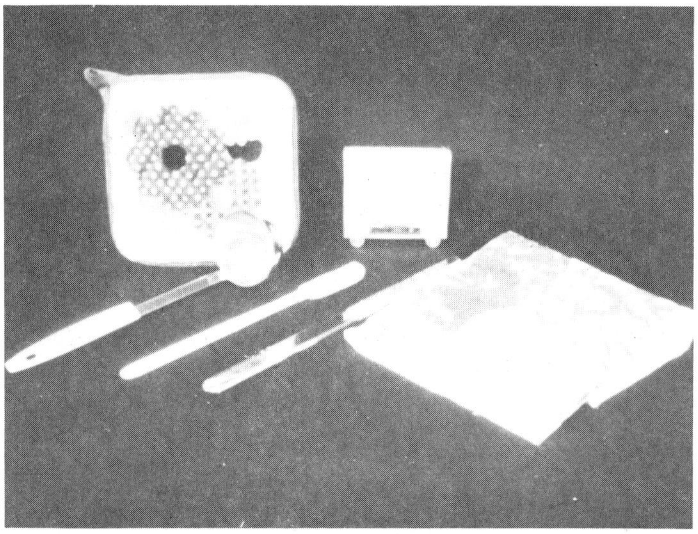

*Figure 6.6. Hot pads, a timer, spatulas of rubber or plastic, and soft, absorbent towels are necessary at several points in the canning procedure.*

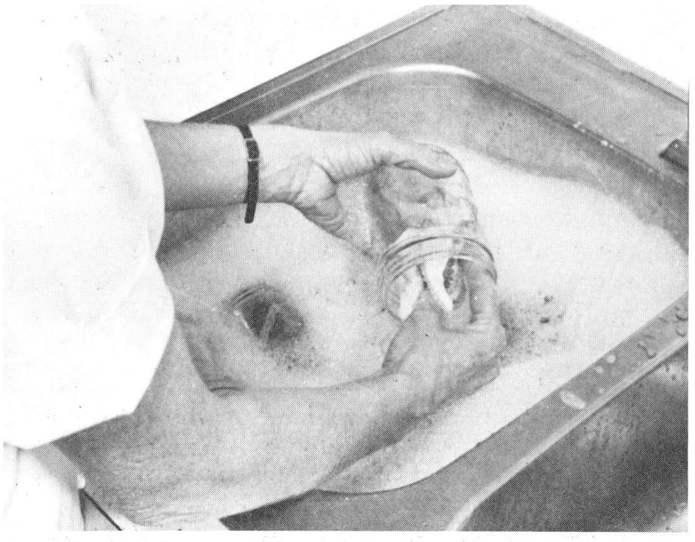

*Figure 6.7. Hot, soapy water is important in preparing clean jars for canning.*

*Figure 6.8. Very thorough rinsing in extremely hot water aids in insuring that the jars are sanitary and ready for the food. Note the very smooth, unmarred lip of the jar, an essential feature if a good seal is to be achieved.*

hot, running water to be certain that no traces of detergent remain. The jars are inverted to drain on clean dish towels. The closures are placed in a shallow pan on the counter. Boiling water is poured over them, and they are allowed to stand in the hot water until needed in the canning process.

The green beans are prepared for canning by washing them well, cutting off the ends, and removing the string. If desired, they may be left whole or cut in the French cut. However, green beans most commonly are cut into pieces an inch long.

**Packing**

Green beans may be canned using either the hot or the raw pack. For the raw pack, the prepared raw beans simply are placed firmly in the jars, salt is added (½ teaspoon for pints, 1 teaspoon for quarts), and boiling water is poured into the jar. The headspace allowed in canning green beans is ½ inch for the beans when they are being packed; the boiling water is added to leave a headspace of ½ inch above the liquid. The lip of the canning jar serves as a useful indicator of this distance.

The *raw pack* method can be used satisfactorily for most vegetables, with the exception of corn, lima beans, and peas. Raw packing is done by first washing the vegetable thoroughly and cutting off any woody or similarly undesirable portions. Then the vegetable is cut into the desired pieces and packed firmly into the glass jars, which have already been washed and rinsed thoroughly. Half a teaspoon of salt is added to each pint jar and a teaspoon of salt is placed in quart jars. Boiling water is added to the jar to cover the vegetable, leaving half an inch of headspace in the jar. The lids are positioned according to manufacturer's directions to complete preparation for processing in the pressure canner.

Vegetables to be *hot packed* are washed and trimmed in the same manner as for raw packing.

*Figure 6.9. After rinsing, the jars are inverted to drain on clean, absorbent towels, where they remain until they are filled with food.*

*Figure 6.10. The closures also are washed and rinsed thoroughly. They are placed in a shallow pan and covered with boiling water until they are used for closing the filled jars.*

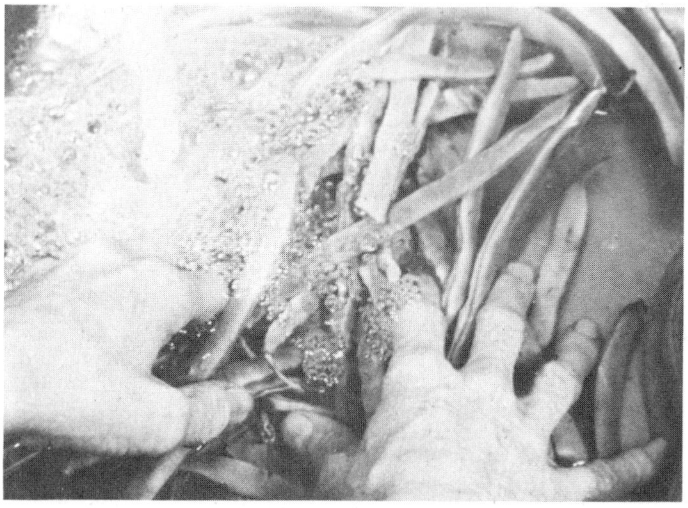

*Figure 6.11. As is true in canning any produce, beans must be washed thoroughly to remove all dirt and other contaminants before being canned.*

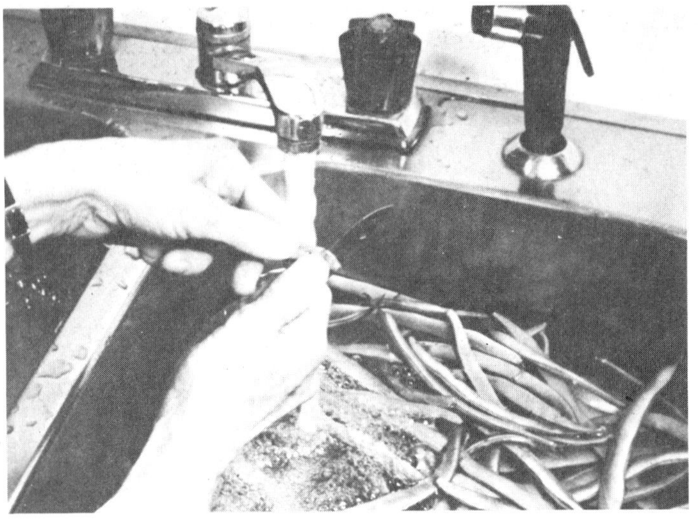

*Figure 6.12. The ends of the beans are cut off, along with any blemishes or spoiled areas.*

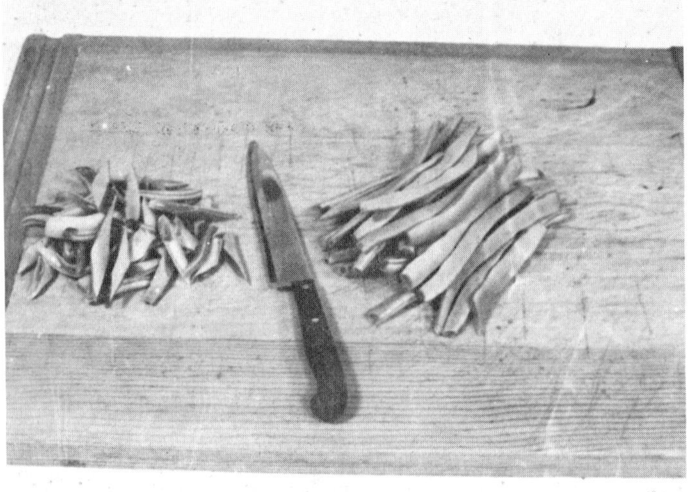

*Figure 6.13. Beans may be cut short or left in longer pieces for canning. The choice is dependent on the probable use of the canned beans.*

*Figure 6.14. Beans being raw packed are placed snugly in the jar, leaving half an inch of headspace in the jar.*

*Figure 6.15. Half a teaspoon of salt is added for a pint of vegetable; a teaspoon is added per quart.*

*Figure 6.16. Boiling water is poured over the raw beans in preparation for canning.*

*Figure 6.17. The headspace for liquid in jars of green beans is half an inch, approximately the depth of the neck of the jar.*

However, the prepared vegetables are placed in a pan of appropriate size, covered with boiling water, and boiled according to the directions indicated in Table 5.7. The amounts of vegetables needed to yield a quart of the canned vegetable are presented in Table 5.2.

**Preparing for Closure**

To insure that there are no air bubbles trapped in the jars, a thin rubber spatula is inserted carefully to shift the beans and release any bubbles. A visual examination around the circumference of the jar helps in being certain that all air is released from the contents of the jar. The rubber spatula is a good utensil to use because it will not damage the jar and will cause a minimum disruption of the beans. Metal utensils have the potential for damaging both the jar and the food; therefore, their use is not recommended.

After the air bubbles have been released, the rim of the jar and the neck should be wiped carefully with a damp, clean cloth to remove any spills. This precaution helps to assure that a good seal will be achieved during canning.

**Closure**

Specific directions for the closure being used should be followed. General suggestions can, however, be made. Tongs are useful in transferring the closure or the rubber ring to the jar. Actual positioning is done using the hands, but it is important that hands be well washed to avoid any contamination of the closure.

The ring of the two-piece lid is screwed on after the center closure is in place. The ring should be tightened firmly by hand. A rubber lid tightener or other tightening device should *not* be applied because the lid will be too tight. These lids are designed to allow some loss of air during processing. This escape of air from the jar is important to proper processing. A very tight lid will inhibit the evacuation of the jar. In addition, the ring will force the closure onto the glass rim too firmly, and the integrity of the ring of sealing compound on the closure may be impaired. This damage to the closure will result in imperfectly sealed jars, thus creating potential spoilage problems.

The porcelain-lined zinc caps are tightened a bit differently, and this distinction needs to be kept in mind. As is true with the two-piece closures, the lid is tightened by hand, not with a tightening device. However, the tightened lid is then turned back a quarter of an inch to facilitate evacuation of air from the jar during processing. It is important to remember that the zinc lids need to be hand tightened again following processing.

With the recent revival of interest in home canning, other closures have and probably will continue to be introduced into the marketplace. The specific instructions for the use of reusable lids and other closures should be noted and followed carefully.

*Figure 6.18. A rubber or plastic spatula is inserted carefully and moved in the jar to help remove any trapped bubbles of air.*

*Figure 6.19. The lip of the jar is wiped absolutely clean with a clean, damp cloth prior to closing the jar.*

*Figure 6.20. Tongs provide a sanitary means of transferring the lid from the hot water to the top of the filled jar.*

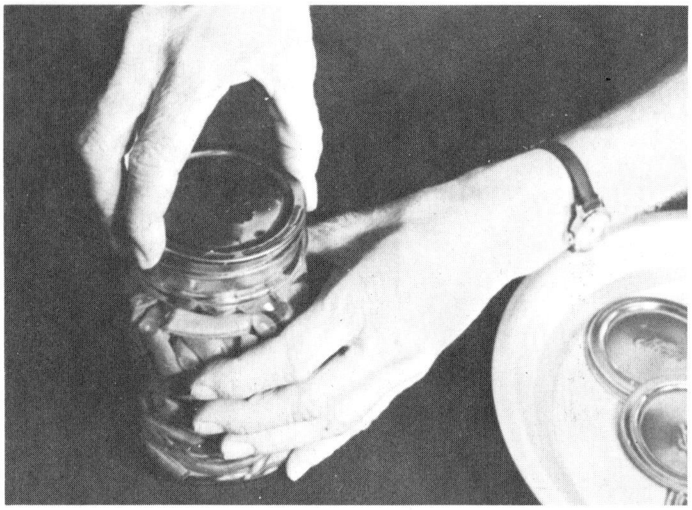

*Figure 6.21. The lid is adjusted carefully with hands which have been washed very clean.*

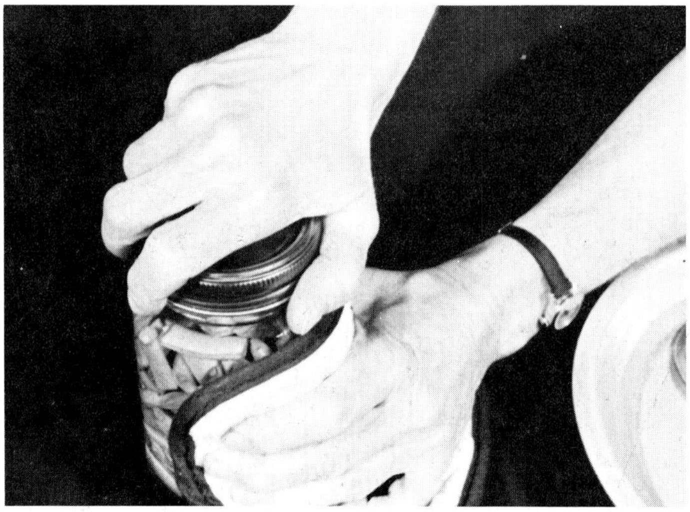

*Figure 6.22. The lids are tightened only with the strength of the hand to avoid damaging the sealing compound in the lid.*

**Processing**

While the beans are being packed into the jars, the pressure canner should be prepared to receive the jars. This is done by placing the rack in the bottom of the canner, placing the canner on the range, and adding boiling water to a depth of an inch for small canners and two to three inches for large canners. The jars are placed on the rack in the canner as soon as each one is ready. Some space should be allowed between each jar to allow for circulation of steam during processing. Only enough jars are packed with beans at this time to fill the canner. Beans which cannot be processed in a single batch should be kept refrigerated until they can be processed in a continuous sequence.

When the canner is filled with the jars of beans, the cover of the canner is locked into position, as determined by its design.

Processing is initiated by heating the canner and its content with the petcock open. Heating is done to exhaust the canner. The time should be noted

*Figure 6.23. A pressure canner is prepared by placing the rack in the canner and then adding water to a depth of two to three inches.*

*Figure 6.24. The jars of food are placed in the canner as each jar is filled, using the jar lifter.*

*Figure 6.25. The cover is placed on the canner as soon as all of the jars are in place on the rack. Note that the jars do not touch each other in the canner.*

*Figure 6.26. The lid of the canner must be locked in position securely for pressure processing.*

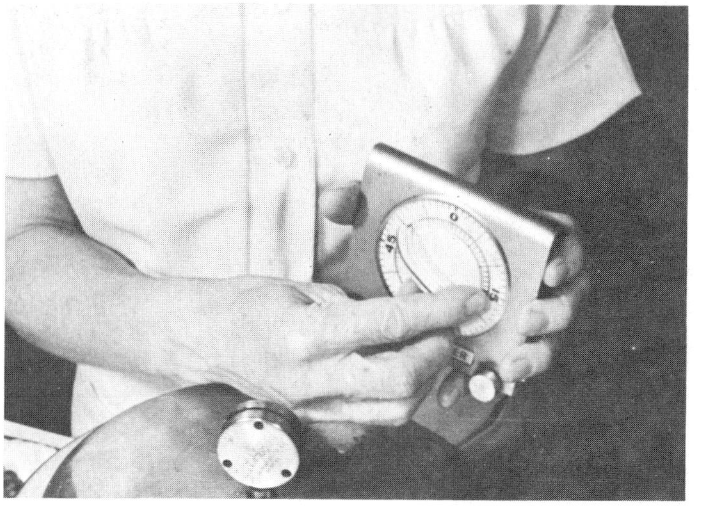

*Figure 6.28. When the weighted gauge begins to jiggle three times per minute, the timer is set for the required processing time, and the heat is adjusted to maintain the jiggling of the gauge about every 20 seconds.*

when steam passes continuously through the opening, and the heating should continue for ten minutes with the petcock open.

Following this preliminary heating period, the petcock is closed immediately to enable the pressure in the canner to build to the desired level (ordinarily to ten pounds). When the dial gauge indicates the correct pressure or when the weighted gauge jiggles, timing for processing begins. For pint jars of green beans, the pressure processing should be continued for 20 minutes for pints and for 25 minutes for quarts. For processing at altitudes above sea level, consult Table 5.8.

The heat needs to be adjusted during processing to maintain the correct pressure. This is done by monitoring the dial gauge to determine that the ten pounds of pressure is being maintained or listening to be certain that the weighted gauge is jiggling at the rate of three or four times each minute.

*Figure 6.29. If the canner has a dial gauge, timing is begun for processing when the gauge indicates the desired (ordinarily 10 pounds) pressure. Note that the petcock at the rear of the cover has been closed so that the pressure will increase in the canner.*

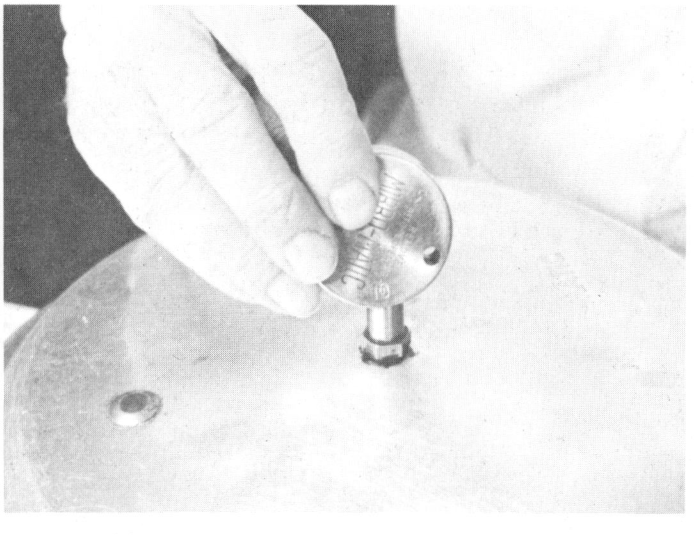

*Figure 6.27. After the canner has been exhausted and steam has been coming continuously from the petcock for ten minutes, the gauge is placed in position, being sure the ten pound indicator is the one being used.*

**Cooling and storage**

When the correct processing time has elapsed, the canner is removed from the heat, but no attempt is made to hasten the cooling of the canner. The canner is left undisturbed until the pressure has returned to normal. At this point, the petcock is tested gently to see if any steam escapes when the petcock is moved. When no steam issues from the vent, the cover can be removed safely from the canner. The lid should be lifted away from the operator so that steam is directed away from the body. With the aid of the jar lifter, the jars now can be lifted from the canner onto a cooling rack or to a pad of folded towels. It is important to place the jars on surfaces of these types to avoid thermal shock, which can happen if the jars are placed directly on a cool surface. Thermal shock can weaken or break the jars at this point. It is also necessary to avoid placing the jars in cold drafts for the same reason. However, there should be adequate space between the jars to allow air to circulate freely around each jar.

Nothing is done to disturb the closures on jars capped with two-piece lids. It is important to avoid tightening the screw band on these jars at this point. The proper closure needs to be achieved prior to processing; tightening following processing can prevent the achievement of a good seal. In contrast, the zinc lids should be tightened by hand immediately upon removal from the canner. This tightening is imperative if the seal is to form. Other types of closures should be treated according to manufacturer's instructions.

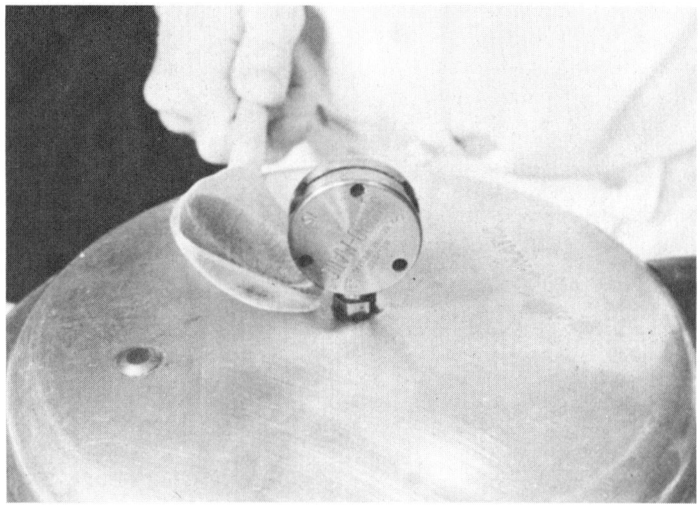

*Figure 6.30. After the processing time has elapsed, the canner is cooled naturally to atmospheric pressure. This state can be checked by gently nudging the weighted gauge to see whether any steam escapes.*

**Testing the seal**

After the jars have cooled completely, they should be examined to be sure they have sealed. With the two-piece closures, the sound of the lids snapping down as the seal forms will have been noted as the jars cool. However, it is necessary to examine each closure individually. One test is to tap the lid with a spoon. A ringing sound is an indication that a seal has been achieved. If any food is touching the lid, a dull sound, yet one which does not sound hollow will be the result when the jar is sealed.

A more reliable test than the auditory one is a visual examination. The lid of a sealed jar should be pulled down in the center, a depression which is sufficient to be observed. A similar check is done by pressing very gently in the center of the lid. A sealed jar will not move downward when pressed, whereas a lid which has not sealed will be flexed downward easily with minimal pressure.

Closures for zinc lids with rubber rings are checked by tilting the jar to see if any liquid leaks from the jar. If there is no sign of leakage, a seal has been achieved.

Jars which have sealed satisfactorily now are ready for storage. The ring is removed from the jars which have been sealed with the two-part closures. The zinc lids are not disturbed in any way prior to storage.

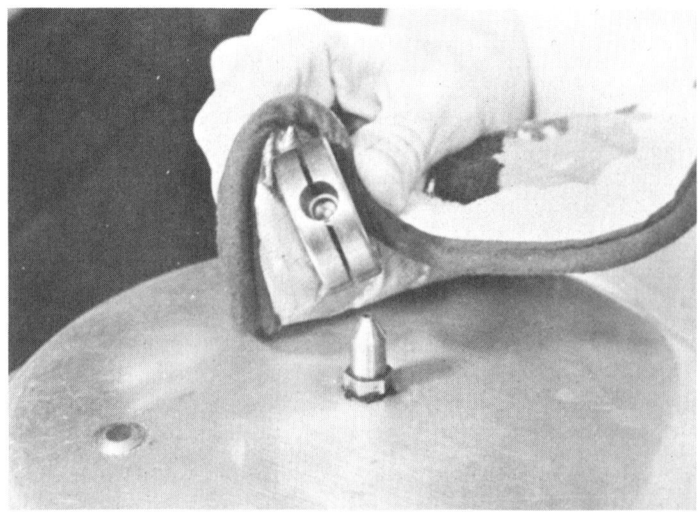

*Figure 6.31. When no steam escapes, the weighted gauge is removed.*

**Labeling**

Although it may seem to be unnecessary, it actually is important to label home canned foods. This is true of any food, including beans. The label should state the date, including the year, very clearly. The supply of canned goods should be rotated at least on an annual basis, but inclusion of the year as well as the day and month is an aid in the event that a few jars may still be remaining in the inventory when the next supply is being canned.

*Figure 6.32. The canner is allowed to cool still more according to manufacturer's directions, and then the cover is removed, being careful to direct the steam away from the body to prevent burns.*

**Storage**

It is apparent that only properly sealed jars can be stored at room temperature. Improperly sealed

*Figure 6.33. The warm jars are lifted from the canner to a rack or absorbent towel with the aid of the jar lifter to finish cooling.*

*Figure 6.34. The finished product is checked for seal, labeled, and placed in a cool, dark place for storage. The shelf life will be excellent for a year, thus representing quite a contrast to the short storage period available to unprocessed produce.*

jars will need to be stored in the refrigerator for prompt use (treated like a fresh food) or else reprocessing will need to be done. The sealed jars can be placed in storage as soon as they have been checked for a seal and have been labeled. Storage ideally is done in a cool, clean, dark place. This is likely to mean storage on the bottom shelf of a storage cabinet in the kitchen. A basement may provide excellent storage if it is available. However, many homes today will have storage available only above ground. Since warm air rises in rooms, the lower the shelf position for storing canned goods, the better will be the result. Canned goods should not be stored in garages or other areas which are not insulated. This is particularly important in severe climates. Canned goods can become frozen and can crack the jars. Hot climates can also cause storage problems if canned goods are stored in garages because of the extreme fluctuations in temperature, and particularly, the overly hot environment in the middle of the day.

Of course, jars should be perfectly clean on the exterior when they are stored. If there has been liquid from the jar overflowing during processing, the jars should be wiped clean with a damp, clean cloth prior to storage. The jars should be kept in covered storage to avoid collecting dust on them.

#### Using canned vegetables

Even with the above procedures, it is important to be cautious about the use of canned vegetables. The contents of the freshly opened jar should be poured directly into a saucepan and heated to a boil. After the boiling point is reached, timing should begin. The beans (or other vegetable) are boiled for 15 minutes without tasting them. This precaution is taken to avoid any possibility of botulism from the food. The subject of botulism is discussed in Chapter 2 in detail.

### Tomatoes

Tomatoes generally are quite acidic, particularly when compared with other vegetables. However, there is some variability in the various varieties, as discussed in the preceding chapter. It is important to be certain that the tomatoes being canned are sufficiently acidic for safe canning without use of the pressure canner. If the tomatoes naturally test at a pH of 4.5 or lower, they can be canned in a water bath canner. For tomatoes with less natural acidity, it still is possible to can these varieties in a water bath canner *if the tomatoes are acidified prior to canning.* Tomatoes can be acidified by adding two teaspoons of lemon juice per quart of tomatoes, two teaspoons of vinegar (at least 5%), or ½ teaspoon of citric acid to each quart. Following addition of the acid, the acidity of the acified tomatoes can be checked conveniently using pH paper.

As was true in the canning of green beans, high standards of sanitation and careful timing of the processing are essential to the preparation of high quality, safe products. However, the processing is less time-consuming than canning with a pressure canner because it is not necessary to let the water bath canner cool as slowly as the pressure canner. Of course, there also is no need to monitor the pressure when canning using a water bath canner. It is essential that the water be maintained at an active boil during the water bath canning operation.

## FRUITS

Fruits are sufficiently acidic to permit them to be canned under somewhat less rigorous conditions than must be employed for canning vegetables. Since it is not necessary to reach temperatures above boiling, a water bath canner can be employed for processing. Use of the water bath canner is a time saver because it is not necessary to exhaust the canner, nor is it necessary to let the canner sit for a long period before opening it. Because of these differences, fruits can be canned with greater

*Figure 6.35. Tomatoes for canning should be ripe, yet reasonably firm, rather than soft. The soil and possible pesticides are scrubbed from the surface as the first step in canning.*

*Figure 6.36. Blanching to loosen tomato skins is a helpful step in preparing a quality product. A wire basket makes blanching an easy task. Note the use of the water bath canner for this operation.*

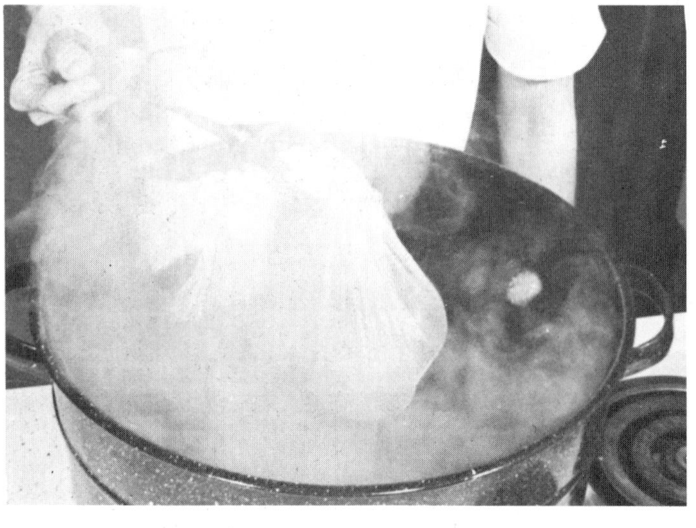

*Figure 6.37. A large piece of cheesecloth can be used very satisfactorily for blanching fruits and tomatoes.*

*Figure 6.38. A quick dip of the blanched tomatoes in ice water loosens the skins.*

*Figure 6.39. The core is removed from the blanched tomatoes.*

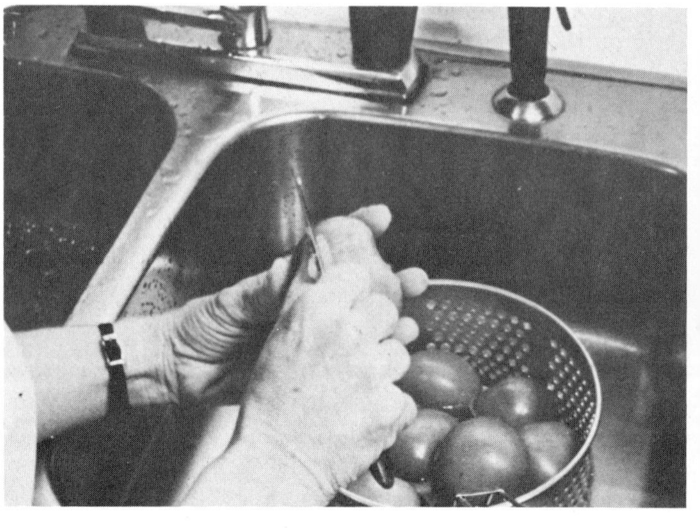

*Figure 6.40. Now the skin can be peeled very easily and efficiently. This is an important means of saving time when quantities of tomatoes are being processed.*

*Figure 6.41. To enhance the flavor of tomatoes, half a teaspoon of salt is added to pints, and a teaspoon is added to quart jars.*

*Figure 6.42. Half a teaspoon of crystalline citric acid can be added to each quart of tomatoes as a means of helping to insure that tomatoes can be processed safely in a water bath.*

*Figure 6.43. Two teaspoons of 5% vinegar (acetic acid) per quart of tomatoes is an alternative means of acidifying tomatoes.*

*Figure 6.44. Yet another acid commonly used to acidify tomatoes is lemon juice (at the rate of two teaspoons per quart of tomatoes).*

*Figure 6.45. More tomatoes may be placed in a jar if the pulp is mashed a bit with a wooden spoon.*

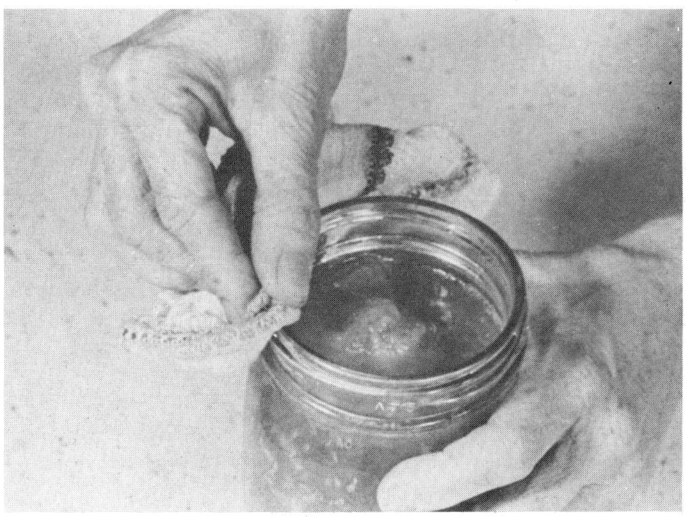

*Figure 6.46. The rim of the jar is wiped clean with a clean cloth to insure a seal can be accomplished.*

*Figure 6.47. Tongs are used to transfer the lid to the jar.*

*Figure 6.48. The lid is adjusted to the correct position with very clean hands.*

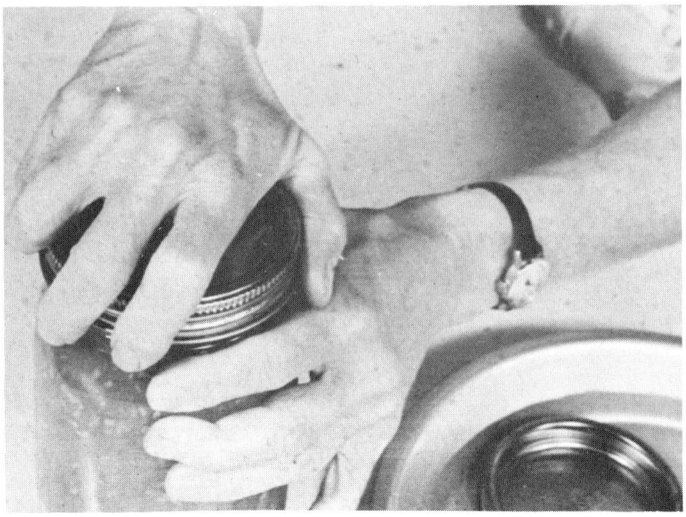

*Figure 6.49. Only the hands are used to fasten the lid securely on the jar.*

*Figure 6.50. As the jars are filled, each one is placed on the rack in the water bath canner.*

*Figure 6.51. Boiling water is poured between the jars. The level of water commonly used in water bath processing is approximately an inch above the top of the jars.*

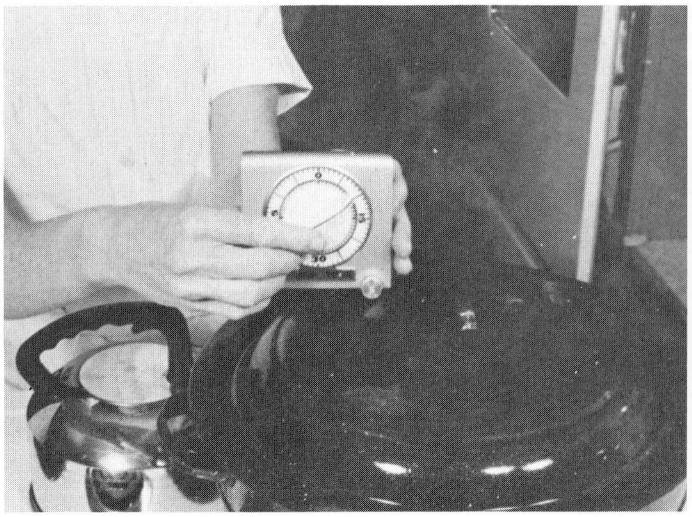

*Figure 6.52. The lid is placed on the water bath canner, the heat is turned up, and the water is brought to a boil before the timer for processing is set. The water is maintained at an active boil throughout the processing period.*

efficiency than vegetables can be processed. The steps involved in water bath canning are reviewed below, using pears as the example.

## Pears

### Preparing the jars

Preparation of the jars and closures is identical to the procedures followed in pressure canning. No modification needs to be made to accommodate water bath canning.

### Preparation of the fruit

Pears, as is also true for other fruits, need to be washed very well prior to being peeled. The pared fruit is cut in halves and then cored. This procedure needs to be done efficiently to minimize the likelihood of discoloration of the pears. Problems with browning can be minimized by immersing the prepared halves in an acidic fruit juice, in an ascorbic acid solution, or in a salt-vinegar solution (1 tablespoon salt and 1 tablespoon vinegar in 2 quarts of water).

### Packing the jars

Either a raw or a hot pack can be used satisfactorily for canning pears. To pack raw, the prepared fruit is placed in the jar, leaving ½ inch of headspace. Pour boiling water or boiling sugar syrup over the fruit, leaving ½ inch of headspace in the jar. Since the sugar content is not high enough to effect preservation of the fruit even when a heavy syrup is used, there is no requirement that a sugar syrup be used. Persons who are canning foods for diabetics or those who are concerned about calorie content may prefer to pack in boiling water.

Pears to be canned by the hot pack method are heated through in hot sugar syrup or in hot water. After the fruit has been packed in the jar to within ½ inch of the top, the cooking liquid is added, leaving ½ inch of headspace. As is true with the raw pack, water or a sugar syrup can be used.

Sugar syrups for canning fruits can be prepared to suit individual preference. A thin syrup is prepared by using a ratio of 1 part sugar to 2 parts water. A medium syrup is based on a ratio of 1½ parts sugar to 2 parts water. Equal amounts of sugar and water (a 1 to 1 ratio) or even an excess of sugar (1¼ parts sugar to 1 part water) are used to prepare a heavy syrup. Enough syrup should be prepared at one time to fill the number of jars which will be processed in the water bath canner in a single load.

### Processing

When the jars have been wiped and sealed, the filled jars are placed in the water bath canner, which contains the rack and water needed for processing. This assemblage should be heating while the jars are being filled so that processing will not be unduly long. The jars should be positioned to allow water to circulate freely around them as well as under them. The rack needs to be designed to permit free water circulation.

After the jars have been placed in the canner, the water is brought to a rolling boil. Then the processing time begins. Pints of pears require a processing time of 25 minutes in the water bath canner, and quarts require 30 minutes. The water in the water bath canner should be maintained at a rolling boil throughout the processing period.

### Treatment following processing

Unlike the precaution required with pressure canners, it is perfectly safe to remove the cover from the water bath canner as soon as the processing time has elapsed and to remove the jars to a rack or towels, as described for vegetables. Zinc lids need to be tightened as soon as the jars have been moved to the cooling area, just as was true for vegetable canning. Other closures should be treated according to manufacturer's directions.

The remainder of the procedures and the treatment during storage are the same as for canned vegetables. The only difference is in the use of the finished product after storage. Fruits, because of their acidity, do not provide the threat of botulism. Consequently, they can be served without being boiled.

## Clear Concord Grape Juice

Yield: 6 quarts

6 cups Concord grapes (without stems)
3 cups sugar
6 cups boiling water

1. Wash grapes very well, being sure to remove all stems. Place 1 cup grapes in a sterilized quart jar. Repeat with the other 5 quarts.
2. Pour ½ cup sugar into each jar.
3. Pour boiling water over the grapes and sugar, leaving ½ inch of headspace.
4. Seal the jars and place on a rack in a water bath canner.
5. Process in actively boiling water for 10 minutes in the water bath canner.

# MEATS, POULTRY, AND FISH

Meats, poultry, and fish present the same potential for botulism that is found with vegetables. Therefore, pressure canning is the method of processing which must be done for safety. Also, the canned product needs to be boiled for 15 minutes before it is even tasted.

### Preparation for canning

Space in jars is at a premium when foods are being prepared for canning. The time involved in processing meats, poultry, and fish is much longer than is required for canning most other foods. In addition, canners hold limited numbers of jars, and the jars represent an investment which cannot be reused until the canned food is consumed. For these reasons, meats ordinarily are removed from the bone prior to being placed in the jar as a means of helping to conserve space. When bones are to be removed, this can be done most easily after the meat has been cooked. However, cooking and canning of meats and poultry should not begin until the carcass has cooled following slaughter. This is necessary to achieve the desired degree of tenderness in the finished product.

### Packing and processing

Meats are packed efficiently into the jars. If salt has not been added to the meat during precooking, ½ teaspoon of salt is added to a pint jar after the meat has been packed fairly firmly. Then a small amount of the cooking liquid (approximately 2 tablespoons per pint) is added. Jars are sealed and processed in the pressure canner in the manner outlined for vegetables. Canned chicken and most meats are processed at 10 pounds of pressure for 75 minutes for pints or 90 minutes for quarts. Fish are processed for 110 minutes for pints. Quarts should not be canned.

After the correct processing time has elapsed, canned meats need to be removed from the canner only after the canner has been cooled without disturbance. Cooling, labeling, and storage are done in the manner described for other canned products.

*Figure 6.53. Chicken is very bulky on the carcass. For canning, it is common practice to stew the chicken until the flesh is tender and can be removed from the bones easily.*

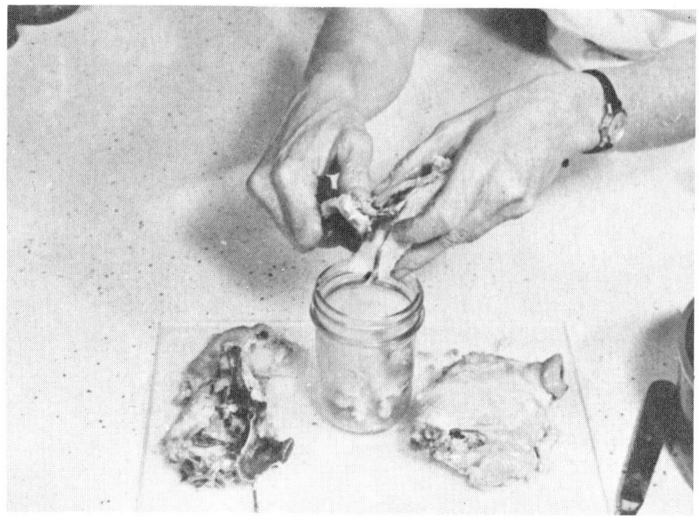

*Figure 6.54. The flesh can be cut from the bones directly into jars. A half pint jar often is a convenient size for canning boned chicken.*

*Figure 6.55. After the jar is filled, the hot stewing liquid is poured into the jar, leaving a headspace of half an inch.*

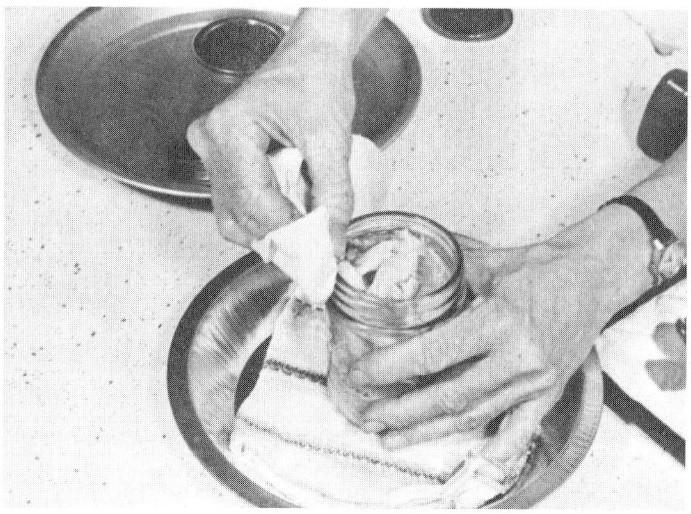

*Figure 6.56. A damp cloth is essential to cleaning the lip of the jar. The canning of chicken requires particular attention to this step because the filling of the jar is somewhat cumbersome, and some flesh may be found on the rim.*

*Figure 6.57. The lid is placed in position with tongs.*

*Figure 6.58. Clean hands are used to adjust the position of the lid.*

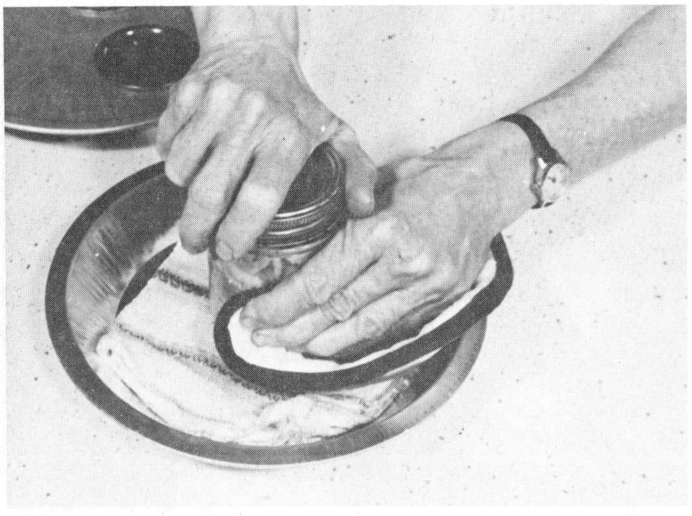

*Figure 6.59. The ring is tightened by hand.*

*Figure 6.60. Jar lifters are very helpful in safely handling hot jars. This half pint jar can be processed easily, along with other small jars of chicken, in a pressure saucepan. For this type of operation, a pressure canner is not necessary.*

*Figure 6.61. Cubes of meat can be packed quickly and efficiently into jars for canning without taking time to cook the meat first.*

*Figure 6.62. Salt is added to the jar of beef for flavor.*

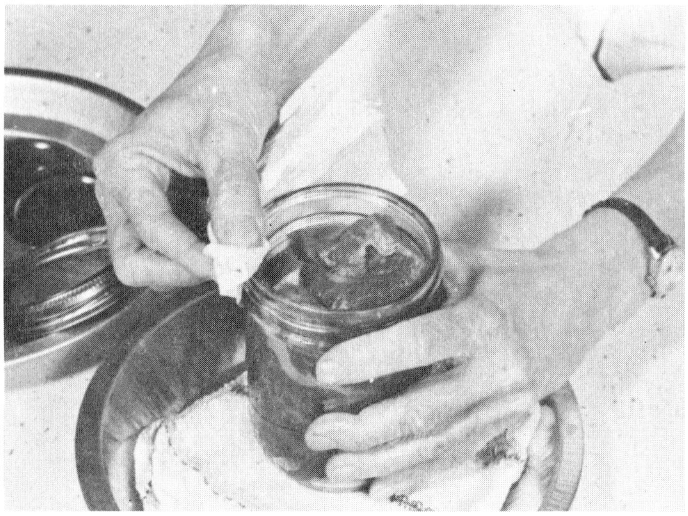

*Figure 6.63. Beef being raw packed is packed with boiling water to leave a headspace of half an inch. Then the rim of the jar is wiped clean.*

*Figure 6.64. The lid is transferred from the hot water bath to the filled jar.*

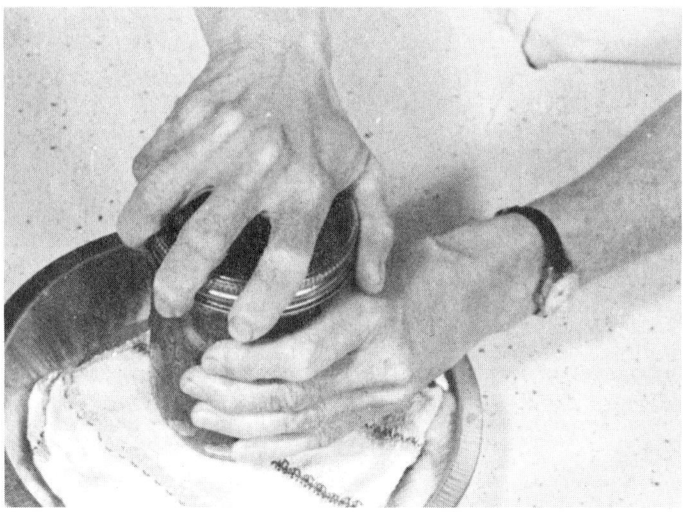

*Figure 6.65. Hand tightening prevents over-exertion of pressure on the sealing compound in the lid.*

*Figure 6.66. Like chicken, small jars of beef are convenient to process and to use.*

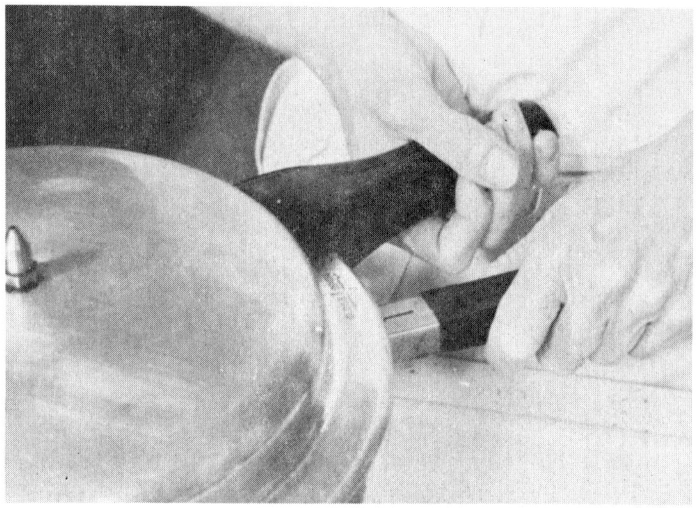

*Figure 6.67. When poultry and meats are being processed in a pressure canner, the lid must be locked securely in position.*

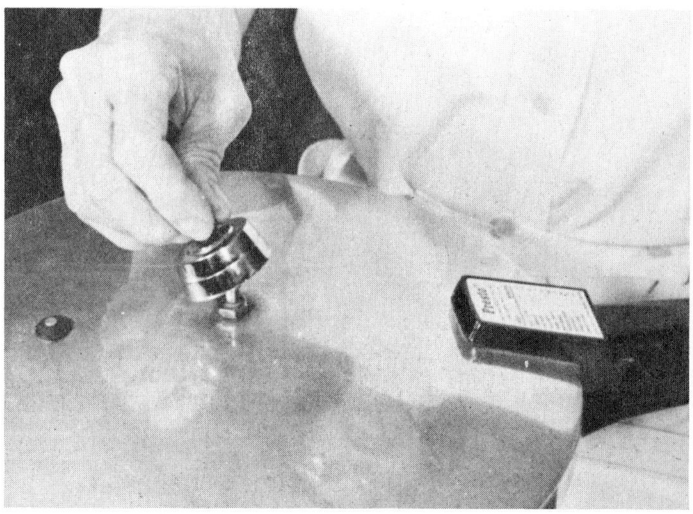

*Figure 6.68. Following manufacturer's directions, the petcock is closed, and pressure will begin to build in the saucepan.*

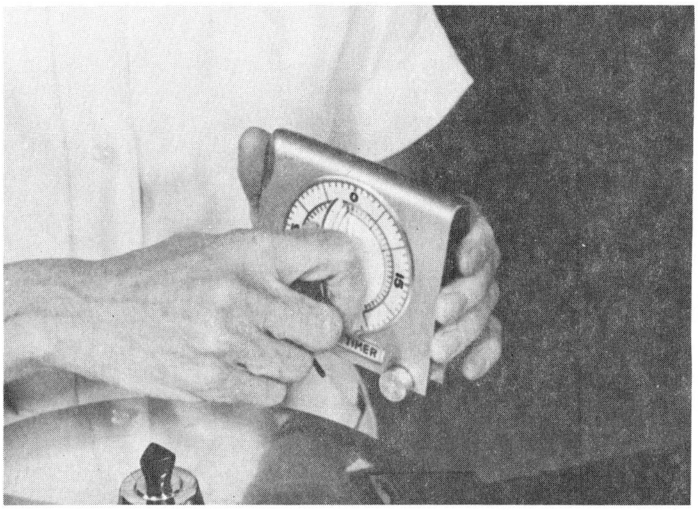

*Figure 6.69. When the desired pressure is reached, the timer for controlling processing is set. When the processing has been completed, the saucepan is cooled gradually, according to manufacturer's directions. This process is somewhat faster than the cooling of the larger pressure canners.*

*Figure 6.70. Fresh fish also can be canned at home.*

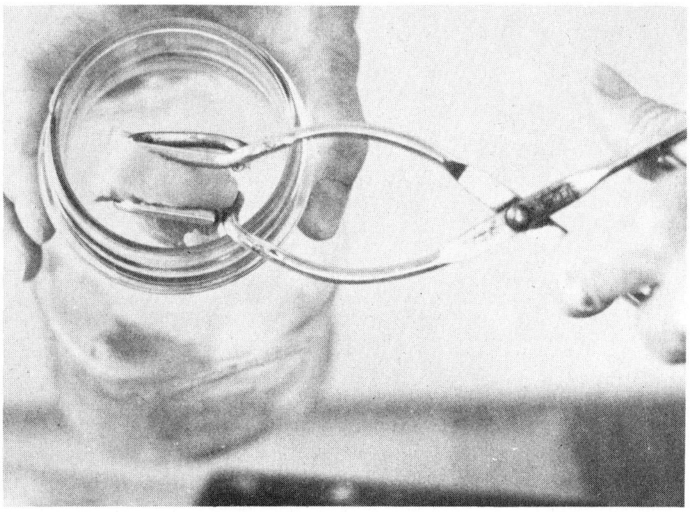

*Figure 6.71. Efficient packing of pieces of fish saves important space in canning jars.*

**Use**

It is imperative that all canned meats, fish, and poultry be boiled actively for 15 minutes prior to even tasting them following canned storage. This precaution is taken to avoid any possible risk of botulism.

# Chapter 7.
# Pickling

## THEORETICAL BASES OF PICKLING

Pickling represents a variation from the canning methods discussed in the two preceding chapters. One difference is found in the acidity of the liquid medium used in canning the product. As has been noted in Chapter 2, the pH of a food is of great significance in determining the processing required for safe preservation of a food. In the case of pickled products, the acidity required is achieved by using vinegar. Another unique feature of pickling is the unusually large amount of salt often added to the product.

### Use of Acids

Pickles are flavorful products which are of importance because they represent a palatable and safe way to preserve food. One of the key factors in their appeal as a food is their flavor interest. Much of this zestiness is contributed by the vinegar which is used to facilitate preservation of the fresh produce. One of the interesting facets of pickling is that vinegar can be used to make vegetable flavors more exciting and also to heighten the interest of some fruit flavors. The wide acceptance of pickles of many types and, indeed, the large amount of shelf space devoted to commercial pickles in the grocery store serve as evidence of the appropriateness of pickling as a means of preserving foods in a palatable form for later consumption.

The vinegar used in pickling needs to be of sufficient acidity to assure that low-acid vegetables can be processed safely under the usual circumstance, that of water bath canning. The vinegar should be 40 or 60 grain, that is 4 to 6 percent acidity. The strength of the vinegar should be indicated on the label. If no indication of strength is found, the vinegar should not be used. Vinegar of 4 to 6 percent acidity should not be diluted. The acidity of pickling solutions needs to be maintained below a pH of 4.5 if water bath canning is to be utilized. If the vinegar were diluted with water, the pH of the medium would rise, with potentially dangerous consequences. Particularly when low-acid vegetables are being pickled, the acidity of the medium is of utmost importance.

The acidic environment of pickle making has another dimension, that of its influence on texture. One characteristic associated with pickles of high quality is their crispness. Soft pickles are considered to be of lesser quality. When fresh vegetables are boiled in an acidic medium, the softening of the cellulose is retarded a great deal. This same phenomenon is observed in the processing of pickles. The presence of the acid needed to facilitate safe heat processing has the added advantage of inhibit-

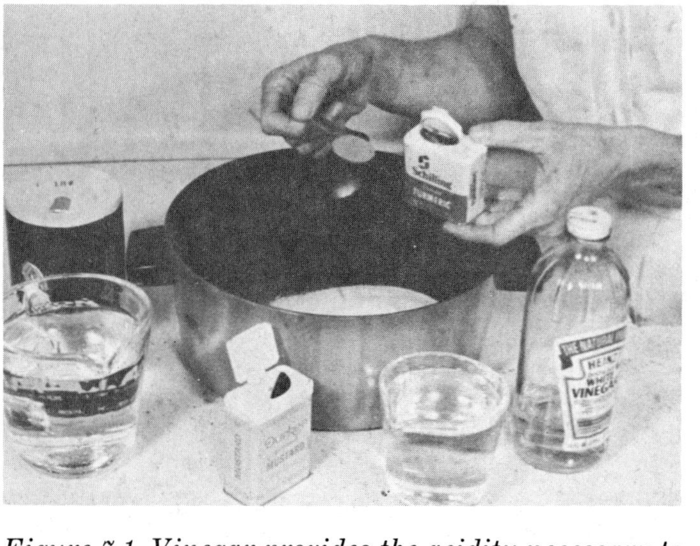

*Figure 7.1. Vinegar provides the acidity necessary to enable pickles to be processed safely by water bath canning.*

ing the softening of the cellulose, thus aiding in the preparation of pickles with the desired crisp texture.

### Salt

Salt has long been recognized as a useful preservative for foods. In high concentrations, salt effectively inhibits growth of microorganisms by creating an unfavorable balance of osmotic pressure between the cellular environment of the microorganisms and the external environment. Salt in high concentrations in a food causes water to be drawn from cells because of the imbalance in the concentrations of the liquids on either side of the semi-permeable membranes of cell walls. This unfavorable balance results in the dehydration of microorganisms, thus promoting safe storage of food.

Heavy brining sometimes is used in preserving foods to provide the main means of halting microorganismal growth in foods such as salted fish. The concentration of the salt brine used in pickling is far less, but the salt content in pickles still is sufficient to enhance their safe preservation.

The use of a very strong brine solution can have an undesirable effect on the texture and appearance of pickles. A very strong salt brine will begin to draw water from the cells of the food being pickled; this is the result of developing an unfavorable osmotic circumstance. If very much liquid is drawn from the plant cells, the dehydrated cells will begin to assume a shriveled appearance because there will not be enough liquid remaining to maintain the desired plump appearance. This is the reason that brine solutions of only moderate strength are used in pickling.

Another facet of the utilization of salt needs to be examined briefly. Table salt, as it is marketed today, contains anti-caking agents which will cause the brine surrounding the pickles to be cloudy.

*Figure 7.2. Salt used for pickling should contain neither anti-caking agents nor iodide. The anti-caking agents result in a cloudy brine solution, and the iodide may darken the pickles.*

The potassium iodide used to iodize salt also has a detrimental effect on pickles because it may cause the pickles to darken in color. Special pickling salts, free of any additives, will be effective in producing pickles of good color with a clear and attractive brine. It should be noted that the additives in salt do not present a hazard; they merely detract from the appearance of the canned product.

## TYPES OF PICKLES

Pickles, regardless of the ingredients from which they are made, are used primarily as a flavor accent in the main course of a meal or as a snack. As a group, they can be characterized as being somewhat sour, interesting in flavor and aroma, and crisp and firm in texture. The common denominator is the use of an acid, specifically vinegar, to promote safe preservation of the food being pickled.

For convenience in reviewing the preservation of food by pickling, pickles can be divided into four groups: (1) relishes, (2) fresh-pack pickles, (3) brined (fermented) pickles, and (4) fruit pickles.

### Relishes

Fruits and vegetables may be used to make a range of pickled products which are utilized as flavor accents, usually being served with the main course of a meal. For example, various chutneys often accompany the serving of Indian curries. Hamburgers gain distinction when highlighted with a spicy chili sauce or piccalilli relish. A colorful corn relish may be served with ham or other meats as a flavorful garnish. Horseradish is a traditional accompaniment to a standing rib roast.

### Fresh-Pack Pickles

A range of vegetables can be pickled satisfactorily using the fresh-pack technique. The significance of fresh-packing is that considerable time is saved in the total procedure of pickling. The vegetable to be pickled is brined for only a few hours before the remainder of the canning process is undertaken. Beet pickles are a favorite example of fresh-pack pickling. Dilled green beans and sweet gherkins are also popular. Dill pickles, made from either whole or sliced cucumbers also can be made by this brief procedure.

### Brined Pickles

The traditional dill pickle is made by a more extended procedure than is used in fresh-packing. Brined pickles are made by fermenting or curing the prepared vegetable in a vinegar-salt brine solution for approximately three weeks. The fermentation aids in developing color, flavor, and translucency. During the course of the fermentation period, the green skin of the cucumber undergoes a color change from the green of chlorophyll to the olive drab color of pheophytin. The interior flesh changes from opaque to translucent, and the cucumber develops the provocative flavor associated with brined dill pickles.

Another universally popular example of brined pickles is that of sauerkraut. This popular food is simply brined cabbage, cut in thin shreds and fermented to the desired tartness.

**Fruit Pickles**

Peach pickles are a particularly popular example of fruit pickles. Others often canned include spiced crab apples, pickled pears, and watermelon pickles. These foods gain much of their appeal from the sweet-sour flavor imbued by the syrup in which they are packed. These types of pickles are excellent as accompaniments to the entree at dinner.

Refrigerator pickles –

# PICKLING

## Equipment

The equipment needed for pickling is much the same as that required for canning. Since the pickling liquid is so acidic, it is possible to do the processing of pickles in a water bath canner rather than requiring the use of a pressure canner. The filling of the jars is aided by the use of a large-mouthed funnel which fits the mouth of the canning jar and by a ladle. A jar lifter is an important utensil to reduce the risk of being burned when handling the hot jars.

A pan of unchipped enameled ware, stainless steel, or aluminum is necessary for heating the acidic liquid used in pickling. Cast iron is not a suitable choice because the acid will react with the iron, causing formation of iron-containing compounds which will be detrimental to the quality of the pickles. Also, if pickles are to be brined or fermented, they will need to be held in a large stoneware, glass, or other non-metallic container which will not interact with the fermenting pickles. The crock needs to be equipped with a plate small enough to fit inside the crock, but large enough to hold all the pickles under the surface of the brine. The plate can be weighted by placing a jar filled with water on top of the plate.

If sauerkraut is to be made, it is necessary to have scales or a household balance to weigh both the cabbage and the salt. Because of the bulkiness of the shredded cabbage, this is the best means of ascertaining that the correct proportions are being used.

The glass jars and closures used for canning pickles should be glass Mason jars of the same type as those used in canning other foods. The jars must be in good condition with no nicks in the lip so that a good seal will be achieved and so the jars will not break during the heat processing.

## Ingredients

**Foods for pickling**

A variety of types of foods may be well suited to pickling. Choices range from fish and eggs through selected vegetables and fruits. Herring is perhaps the type of fish most commonly pickled, but other white fish also may be used. It is essential that fish to be pickled must be very fresh and firm.

Cucumbers for pickling should be selected to provide the size of pickle desired. They should be of uniform size. The cucumbers also need to be firm. It is essential that the exterior surface of the cucumbers not be coated with a coat of wax because this coating will prevent the brine from penetrating satisfactorily into the cucumber.

Other vegetables commonly pickled include corn, green tomatoes, onions, green beans, and beets. Corn for pickling as a relish should be from fresh ears of medium size, or it may be preferable to utilize frozen sweet corn, depending upon the season of the year. Although large white onions may be utilized effectively in preparing pickled relishes, only the very smallest boiling onions are well suited to pickling as whole, pickled onions. Such onions need to be firm and free of mold or soft areas. Green beans should be fresh and crisp, with no suggestion of limpness. They also are free of rust or blemishes. Beets for pickling should be young and fresh (about an inch to an inch and a half in diameter); it is important that there not be any suggestion of woodiness, as can occur in the larger, more mature beets.

Best results are obtained when fruits selected for pickling are just slightly underripe. They should just be beginning to yield to pressure. This degree of firmness enables the fruits to undergo the pickling process without losing their shape. The flavor also is developed sufficiently for the characteristics required in making fruit pickles with exciting

*Figure 7.3. Although cucumbers are most often used, many other vegetables, including onions, tomatoes, and green peppers, are utilized in making pickles and relishes.*

flavors. Peaches, pears, apricots, and crab apples are examples of fruits which are well suited to pickling. Watermelon provides another illustration, but it is somewhat unique in that the rind is the portion which is pickled, and the meat of the melon can be consumed fresh, making this a particularly economical type of pickle. Watermelon pickles are best when made from a watermelon with a rind at least an inch thick. The rind should be unscarred and firm.

Foods to be pickled should be refrigerated until they are being prepared for pickling. This will assure that maximum quality is retained in the food. Pickles cannot be better than the food used to make them.

**Other ingredients**

Salt especially prepared for pickling is an important ingredient in the preparation of pickles. The requirements for performance in pickling are quite different from the nutritional considerations utilized in selecting salt for the table and for food preparation. The problems to be considered in selecting salt for pickling are the effect of the salt on pickle color and on clarity of the pickling liquid. Iodized salt is not well suited to the preparation of pickles because of its darkening effect on pickle color. Anti-caking compounds in regular table salt contribute to a cloudy liquid rather than promoting the desired clarity.

The choice of vinegar is influenced by the color of the food being pickled. If pickling onions, the color of white onions would be poor if cider vinegar were to be used, but white distilled vinegar will result in a very satisfactory product. In products where the darker color of cider vinegar is not a concern, this vinegar contributes to a pleasingly mellow flavor. For special flavor effects, other special vinegars may be preferred. Wine vinegars are examples of specialized products to contribute unique flavors.

Even more important than the flavor and color of the vinegar selected is the acidity of the product. The concentration of vinegar is indicated on the label as a percentage acidity or in grains. If acidity is indicated in percentages, the vinegar used for pickling must be at least 4% and often will range as high as 6%. If measured in grains, the appropriate vinegars will range between 40 and 60 grains (figures comparable to the 4 to 6%). These vinegars must be used in the concentrations provided in the bottles. It is essential that vinegars used for pickling not be diluted greatly or the dilution will raise the pH of the liquid and may make the product have such a low acid content that the pickles cannot be processed safely by the water bath canning method ordinarily used for canning pickles.

An ingredient found in some pickle recipes is sugar. When deciding upon the type of sugar to be used, the desired color of the finished product needs to be considered. Although granulated sugar is a very satisfactory choice, the darker color provided by using brown sugar may be preferred.

*Figure 7.4. Fresh spices are important in preparing pickles with optimal flavors. The spices and vinegar used in pickle making add greatly to the flavor appeal of pickles.*

Specific recipes suggest a wide range of spices. For best results in canning pickles, fresh spices should be used. Some of the important flavoring constituents are volatile and, therefore, are lost during shelf storage after the package is opened. If spices must be stored, their useful shelf life can be extended by placing the packages in plastic bags which are sealed tightly against loss of volatiles.

## Preparation for Processing

**Brining**

Some pickles, notably dill pickles and sauerkraut, require a fermentation period of three weeks or more. Fermentation is done by soaking the cucumbers or cabbage in a salt solution of an appropriate strength to permit the desired changes to develop in the product prior to processing. The concentration of salt in the brine is important to the fermentation process and needs to be controlled carefully. It is also important that all of the food being pickled be completely submerged in the brine so that there will not be variations in the product in different parts of the batch. Therefore, the food is immersed completely in the brine by placing a plate or plastic bag filled with water over the surface of the fermenting mixture. The plate (which is weighted with a glass of water or other suitable device) holds the food beneath the surface of the brine and also effectively blocks evaporation of the water so that the brine concentration remains essentially unchanged during the fermentation period. Similarly, a large plastic bag designed for packaging food can be filled with enough water to push the bag out to fill the top of the crock (to prevent evaporation) and to make the bag heavy enough to push the floating food beneath the surface.

The fermentation process is allowed to proceed at room temperature. Gas bubbles forming are an indication that the desired changes are developing in the product. The cover provided by the plate or plastic bag will help to prevent the growth of undesired molds and yeasts. However, a scum will begin to form on the crock in which dill pickles are being fermented after only a few days. This scum can be skimmed off daily during the remainder of the three-week brining period.

Throughout the fermentation period, it is essential to have the food covered with the brine completely at all times. To assure that the level of brine is sufficient and that the cover is holding the food submerged, a daily check is needed. If too much water has evaporated and the brine solution is too low to cover the food, more brine should be added. The added brine needs to be of the same concentration as the original brine. If a plastic bag has been used and it happens to break, the brine solution in the crock should be drained off completely and a new brine solution of the original strength should be added. This is very important because the diluted brine resulting from the addition of water from the bag would make the brine too weak to effect the desired pickling action.

### Other preliminary treatments

Many pickled products do not undergo the prolonged fermentation period utilized for sauerkraut and dill pickles. In fact, dill pickles can even be prepared by a procedure referred to as fresh pack, in which the cucumbers are placed in a brine solution just overnight. Pickle slices may be brined for as short a time as three hours.

Other pickles may be made without any fermentation period. This is true for various fruit pickles, as well as beets and sweet pickles. The pickling of fruits and beets is done in a continuous process with no delay for special treatment. However, sweet pickles undergo a boiling water treatment on three successive days, combined with a limited opportunity for fermentation on the second day, a marinade on the third day, and on the fourth day, which is also the processing day.

## Processing

Pickles are removed from the brine if they have undergone even a short fermentation period. The food, whether being packed fresh or following fermentation, is packed into clean, hot jars appropriate for canning, leaving a headspace of ½ inch. The appropriate liquid (vinegar and other ingredients, according to the specific recipe) is added to leave ½ inch of headspace. The lids are placed on the jars according to the manufacturer's directions, and the jars are ready for processing in the water bath canner.

Pickles are processed in the water bath canner for the length of time recommended in specific recipes. As was discussed in Chapter 5, the timing in water bath processing begins after the water in the canner is boiling actively with the jars in

*Figure 7.5. Dill pickles are an example of a pickle which is brined (fermented in vinegar) for an extended period of time.*

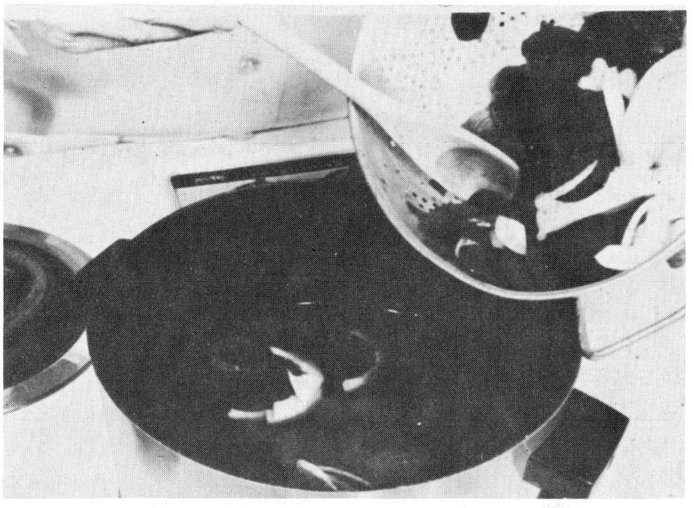

*Figure 7.6. Beet pickles are an example of a pickle which does not undergo fermentation.*

*Figure 7.7. The equipment used in processing pickles is the same as is used for other water bath canning. The funnel, ladle, and jar lifter are particularly useful pieces of small equipment.*

*Figue 7.8. Because of their high acidity, pickles can be processed in a water bath canner. The usual time for processing pickles is ten minutes after the water in the canner is boiling actively.*

*Figure 7.9. Following processing, the hot jars are transferred to a rack where they are cooled away from drafts and with some space between each jar.*

position. Careful timing is important, despite the presence of acid, because yeasts and molds which may be present in the jars must be destroyed during heat processing to avoid spoilage of the pickles during storage.

Following the water bath treatment, jars are removed from the canner and cooled carefully. Zinc lids are tightened immediately upon removal from the canner, while other closures are not disturbed.

Jars are checked for sealing prior to storage. Any jars failing to seal will need to be reprocessed or else refrigerated for use in the near future. Labeling of jars should be done, and then the jars are stored for use, preferably within the next year.

The above steps in processing pickles for long-term storage are presented in an abbreviated form simply as a review. The detailed discussion of the rationale and precautions to observe in each step is presented in Chapter 5 and should be borne in mind during the study of pickling. Pickles, like all other canned foods, have the potential for spoiling during storage. Adequate heat treatment is provided when boiling-water-bath canning is the method of processing utilized, but all of the precautions for sanitation and sealing required in canning fruits also need to be observed in canning pickles.

## EVALUATION

Pickles should be evaluated on the basis of their texture, appearance (including color), and flavor. Texture of pickles is a particularly important criterion of quality. A crisp, meaty pickle is the desired standard. Sometimes pickles may be shriveled, hollow, or soft and slippery. The causes for these textural variations are related primarily to errors in the brining and processing steps, but hollow pickles may be due to using poorly developed cucumbers or to holding the food too long before processing.

Shriveling can also be due to too long a delay from the time the food is picked until it is processed, particularly if the holding period has been in a warm room rather than in the refrigerator. The use of too much salt in the brine solution or too much sugar in the syrup will create unfavorable osmotic pressure, thus causing shriveling even if the food was plump and firm prior to incorporation with the solution.

Softness or slipperiness develops in pickles when opportunities for growth of yeasts and molds are not controlled adequately. For example, the presence of cucumber blossoms provides the opportunity for introduction of the enzymes which will cause softening of the cucumbers during the fermentation period. The use of too weak a salt solution during fermentation or of too weak a vinegar in the packing liquid can also permit excessive growth of molds and yeasts, resulting in soft and slippery products. Very hard water, because of its alkaline nature, may also create too high a pH in the liquid, creating a situation comparable to having too little vinegar. It also is important to remove the scum from brine solutions daily and to keep the cucumbers completely immersed in the brine if fermentation is to proceed normally without too high a yeast population.

Pickles are expected to have somewhat bright, appealing colors. Sometimes they may appear to be darker than desired and to detract from the appeal of the canned pickles. Iron, whether in the water or derived from contact with iron cooking equipment, is one cause of dark pickles. Copper, brass, and galvanized iron are other possible sources of contamination which will cause darkening. If ground, dark-colored spices are used in pickling, these cannot be removed from the product and, consequently, will cause darkening. Use of cider or other dark vinegar will cause darkening in white or light-colored pickles. Iodized salt, if used in preparing pickles, also contributes to darkening. This effect is due to the potassium iodide in the salt.

Dull and faded colors in pickles are related to poor selection of foods for pickling. Cucumbers which are quite yellow in tone will be faded in appearance when pickled. Similarly, peaches or crab apples with poor color will not be attractive in the completed pickles. The same observation can be made for other foods being pickled.

Cloudiness or sedimentation in the liquid in the jar may be the result of using table salt with anti-caking agents. Use of unstrained brine from dill pickles will also cause clouding.

Flavor of pickles may be deemed by some to be too acidic. If this is the case, subsequent pickling should be done by increasing the amount of sugar in the recipe. It is essential that the vinegar be used at the recommended concentration of 4 to 6%. If salt is judged to be too high in the pickles, the brine solution can be washed from the pickles before they are packed. The actual flavor of the food is related most importantly to the quality of the food selected for pickling and the speed exercised in initiating the pickling process following harvesting.

## SUMMARY

Pickles may be classified as relishes, fresh-pack pickles, brined pickles, and fruit pickles. The pre-preparation ranges from a prolonged fermentation period for brined pickles, through very limited fermentation in fresh-pack pickles, to no fermentation in fruit pickles.

The unique textural and flavor qualities in pickles are achieved through the use of vinegar and seasonings of various types; salt often is used to modify flavor and texture, and sugar serves a similar function in fruit pickles. Because of the acidity of the vinegar, pickles can be processed safely by the boiling-water-bath canning method. However, careful attention to the same principles used in canning fruits should also be directed to the canning of pickles.

Evaluation of pickles is done on the basis of texture, appearance, and flavor. Specific deviations detracting from pickles may include soft, slippery, shriveled, or hollow texture; dark, faded or dull colors; cloudiness or sediment in the liquid; and too acidic or too salty a flavor.

## BIBLIOGRAPHY

Ball Corporation. *Blue Book.* Ball Corp., Muncie, Indiana. 29th ed. 1972.

Kerr Glass Manufacturing Corp. *Home Canning and Freezing Book.* Kerr Glass Manufacturing Corp. Sand Springs, Oklahoma. 10th ed. 1975.

Human Nutrition Research Division. *Making Pickles and Relishes at Home.* Agricultural Research Service, U.S. Dept. of Agriculture. Washington, D.C. 1966.

# Chapter 8. Recipes for Pickles

The procedures and principles of pickling were presented in Chapter 7. This chapter contains examples of the various types of pickles, that is relishes, fresh-pack pickles, brine pickles, and fruit pickles. Since so many variations exist in the actual preparation of specific types of pickles, the information needed is given specifically in each type of pickle recipe rather than being presented in a general method.

## RELISHES

### Zucchini Relish

10 medium zucchini (about 10 inches long)
2 large onions
1 large green bell pepper (or 1 small green and 1 small red bell pepper)
¼ cup canning salt
6 cups sugar
2 cups cider vinegar
1½ cups water
1 teaspoon pepper
1 teaspoon turmeric
1 teaspoon nutmeg
½ teaspoon allspice
½ teaspoon ginger
2 teaspoons celery seed

1. Coarsely grind (in a blender) or chop the zucchini, onion, and pepper.
2. If water is used to facilitate grinding in the blender, drain the vegetables well in a colander.
3. Combine chopped vegetables with the salt until well blended.
4. Cover with cold water and let stand overnight at room temperature.
5. Drain the mixture in a colander and then rinse well in the colander under a stream of cold water.
6. Drain well.
7. Combine the sugar and remaining ingredients and heat to boiling in a kettle.
8. Add drained vegetables to boiling liquid and heat to boiling.
9. Simmer the relish for 10 minutes.
10. Pack in clean, hot jars and proceed as for water bath canning. Processing time in boiling water is 10 minutes for pints.

*Note:* Relish may be served with meats, such as hamburgers or may be mixed with salad dressings to make a tartar sauce or a salad dressing variation for use on tossed salads.

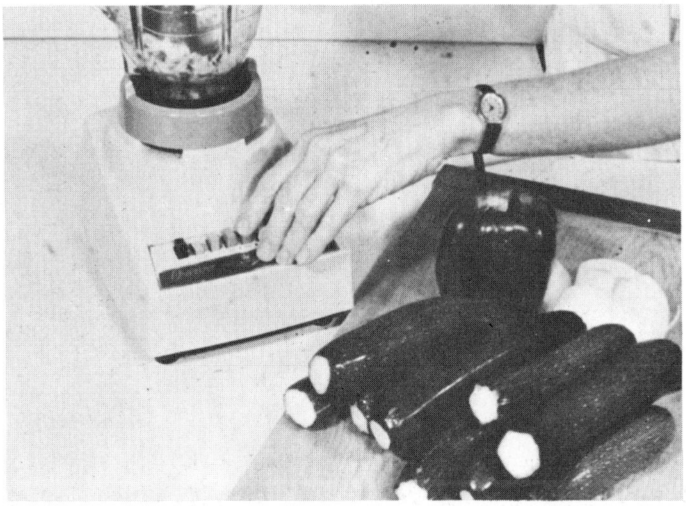

*Figure 8.1. Zucchini, onions, and green pepper are the basis for Zucchini Relish.*

*Figure 8.2.* Considerable time can be saved in preparing relishes by using a blender for chopping.

*Figure 8.3.* The ingredients are whirled in the blender until coarsely ground. The addition of a small amount of water will facilitate the process.

*Figure 8.4.* The chopped ingredients are strained, with the aid of a colander, to remove the excess water which was added in the blender.

*Figure 8.5.* The canning salt is stirred into the ground vegetables, water is added to cover, and the mixture is allowed to ferment overnight at room temperature.

*Figure 8.6.* While the sugar, vinegar, and spices are being heated to a boil, the fermented vegetables are washed thoroughly in a colander under a stream of cold, running water to remove the salt brine. Then the vegetables are added to the hot liquid.

*Figure 8.7.* The relish is simmered for 10 minutes to develop the flavor and enhance the texture of the relish.

*Figure 8.8. Ordinarily the zucchini relish is ladled into canning jars in preparation for processing in a water bath canner. However, short-term storage can be done in the refrigerator in tightly-covered plastic containers.*

## Dilled Onion Slices

½ cup sugar
2 teaspoons salt
½ teaspoon dill weed
½ cup white vinegar
¼ cup water
1 tablespoon sugar
2 large white onions, sliced thinly

1. Boil all ingredients except the onions together for one minute.
2. Cool before pouring over the onions.
3. Pack in clean, hot canning jar. Process in boiling water for 10 minutes (in water bath canner).

## One-Eyed Widow Chow-Chow

1 quart chopped cucumbers (unpeeled)
1 quart chopped onions (peeled)
1 large bunch of celery, chopped fine
2 red bell peppers, chopped
½ cup salt
Boiling water to cover
2 pounds brown sugar
1 box mustard seed
1 tablespoon turmeric
¾ cup sifted flour
1 quart cider vinegar
1 pint (or less) cocktail onions, drained

1. In a large bowl, sprinkle all the vegetables with salt.
2. Cover with boiling water and let stand at room temperature overnight.
3. Thoroughly blend the brown sugar, mustard seed, and turmeric with the flour.
4. Add the vinegar gradually with stirring.
5. Heat to boiling to gelatinize the starch, while stirring constantly to prevent lumping.
6. Drain the vegetables thoroughly.
7. Add the vegetables and the cocktail onions to the vinegar mixture.
8. Heat to boiling. Pack into canning jars, leaving ½ inch of headspace.
9. Boil pint jars in water bath canner for 15 minutes.

## Corn Relish

2 quarts whole kernel corn (16 to 20 ears)
5 sweet red peppers, diced
4 green bell peppers, diced
1 large bunch celery, chopped
¾ pound white onions, chopped
1½ cups sugar
1 quart vinegar
2 tablespoons salt
2 teaspoons celery seed
2 tablespoons dry mustard
1 teaspoon turmeric

1. Husk fresh corn and boil ears for 5 minutes before plunging into cold water. Six 10-oz. packages of frozen corn may be thawed at room temperature or by alternate means and substituted for the fresh.
2. In a kettle combine peppers, celery, onions, sugar, vinegar, salt, and celery seed; heat to boiling and continue to boil uncovered for 5 minutes, stirring occasionally.
3. Thoroughly mix the dry mustard and turmeric before blending it well with some of the boiling liquid.
4. Cut the fresh corn from the cob; add the corn kernels with the mustard-turmeric mixture to the liquid.
5. Boil the mixture another 5 minutes after the corn and spice mixture has been brought to a boil again.
6. If desired, thicken the relish by adding a slurry of ¼ cup flour and ½ cup water when the corn is added to the kettle. This mixture must be stirred almost continuously for uniform gelatinization and minimum opportunity for scorching.
7. Pack relish into pint jars, leaving ½ inch of headspace.
8. Process for 15 minutes after the jars have been placed in a water bath canner and the water in the canner is boiling again.
9. Complete the canning procedures as described in Chapter 6.

# FRESH-PACK PICKLES

## Versatile Pickles

1¼ cups sugar
1 cup vinegar
2 cups water
1 teaspoon salt
1 teaspoon turmeric
1 teaspoon dry mustard
4 large cucumbers
1 large onion
1 small jar of pimiento strips (optional, for color)

1. Place sugar, vinegar, water, salt, turmeric, and dry mustard in a large saucepan and bring to a boil.
2. Peel and slice the cucumbers and onion.
3. Drain the pimiento very thoroughly.
4. Add the cucumbers, onion, and pimiento to the boiling liquid and maintain a rolling boil for 10 minutes.
5. Pack in hot, clean jars; process for 10 minutes in a water bath canner, as described in Chapter 6. (These pickles may be stored in the refrigerator in tightly covered plastic containers for as long as three weeks).

*Note:* Zucchini may be substituted for cucumbers in this recipe, or half cucumbers and half zucchini may be used.

*Figure 8.9. Versatile Pickles can be made using equal amounts of zucchini and cucumbers or only zucchini or only cucumbers.*

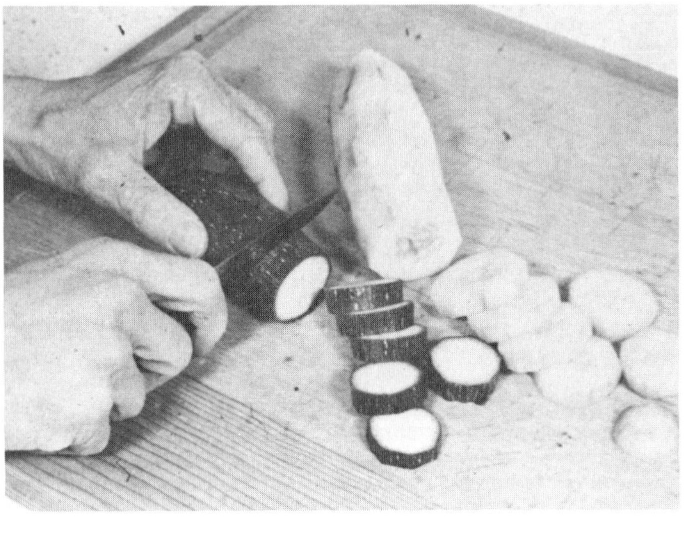

*Figure 8.10. Although zucchini and cucumbers often are peeled for pickles, the tender skin of the zucchini can be left on the vegetable to add color to the finished product.*

*Figure 8.11. The ingredients are boiled together for ten minutes to aid in blending the flavors and modifying the texture. Turmeric, a brilliantly colored ingredient in this recipe, needs to be handled carefully to avoid its intense stains.*

## Curry Slices

8 cups (2½ pounds) thinly sliced, medium cucumbers
1 medium onion, thinly sliced
1 tablespoon salt
2 cups vinegar
1⅓ cups sugar
1 tablespoon curry powder
2 teaspoons pickling spices
1 teaspoon celery seed
1 teaspoon mustard seed
½ teaspoon pepper
1 green pepper, cut in narrow strips

1. In a large bowl, arrange cucumber slices and sliced onions and sprinkle with salt.
2. Cover with ice water, and let stand for 3 hours.
3. Drain vegetables well.
4. Combine remaining ingredients in a kettle, along with the cucumber and onion.
5. Heat just to boiling and then pack into pint jars.
6. Place jars in water bath canner and boil for 10 minutes.
7. Finish canning as described in Chapter 6.

Yield: 3 pints

## Beet Pickles

13 pounds small, young beets
2 cups sugar
2 cups water
2 cups cider vinegar
1 teaspoon cloves
1 teaspoon allspice
1 tablespoon cinnamon

1. Wash beets well; remove all but about an inch of the top and two inches of root.

*Figure 8.12. Pickling cucumbers are sliced thinly or in slices up to half an inch thick to suit individual preference in preparing Curry Slices*

*Figure 8.13. Curry Slices are an example of a fresh-pack pickle. These are heated in a spiced vinegar solution just to boiling.*

*Figure 8.14. The hot, curried slices are packed immediately for processing in a water bath canner.*

2. Cover beets in saucepan with water and boil about 15 minutes until skins slip easily.
3. Plunge beets into ice water to cool so that skins can be removed easily and the roots and stems cut off. Slice the beets.
4. Prepare syrup using the remaining ingredients. Heat to boiling while packing sliced beets into hot, clean jars, leaving ½ inch of headspace.
5. Pour hot syrup over the beets to within ½ inch of the top of the jar.
6. Close jars and process in the water bath canner. Boil for 30 minutes before cooling and proceeding with the remaining steps in canning (checking for seal, labeling, and storing).

## BRINE PICKLES

### Pickled Onions

6 cups white pickling onions
5½ cups water
½ pound iodine-free salt
3 cups white vinegar
3 cups sugar
3 teaspoons pickling spice

1. Wash and peel onions. Place in a crock and cover with the water and salt solution to a depth of at least 5 inches above the onions.
2. Let stand in brine (covered with plate or plastic bag, as described in dill pickles) for 5 weeks.
3. Remove from brine and cover onions with boiling water.
4. After standing in water for 2 hours, drain the onions and pack in hot, clean pint or half pint jars.
5. Heat the sugar and vinegar to a boil and pour over packed onions, leaving ½ inch headspace.
6. Save any extra syrup. Cover jars and let stand at room temperature for a day.
7. At the end of the first day, drain off the syrup from the jars, reheat it to boiling and pour again over the onions. Cover jars and let stand a second day.
8. Again drain off the syrup, reheat it to boiling, pour over the onions, cover, and store at room temperature for a third day.
9. Drain off the syrup and reheat a final time. Pour over onions in the jars, add the pickling spices (distributed uniformly in the jars); close the jars for canning and process in boiling water for 5 minutes in water bath canner. Proceed through checking for seal, labeling, and storing steps.

*Note:* Onions may be prepared easily by pouring boiling water over them and immediately draining them, then transferring into a pan of cold water. If the onions are peeled and sliced underwater, lacrimation will be avoided.

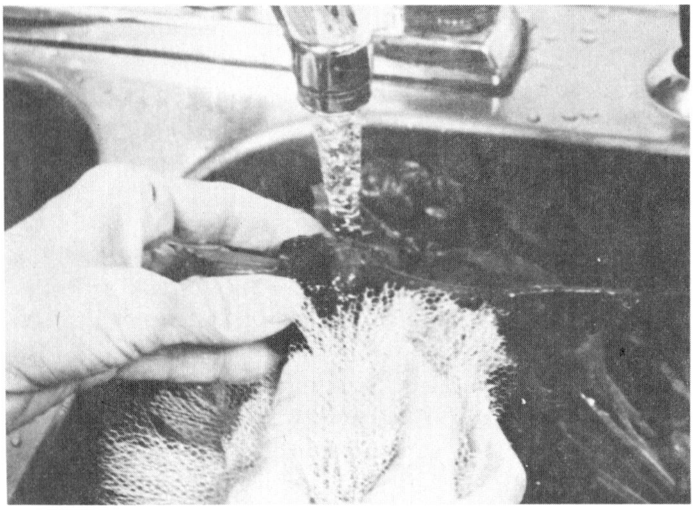

*Figure 8.15. Beets for pickling are scrubbed thoroughly as their first preparation step.*

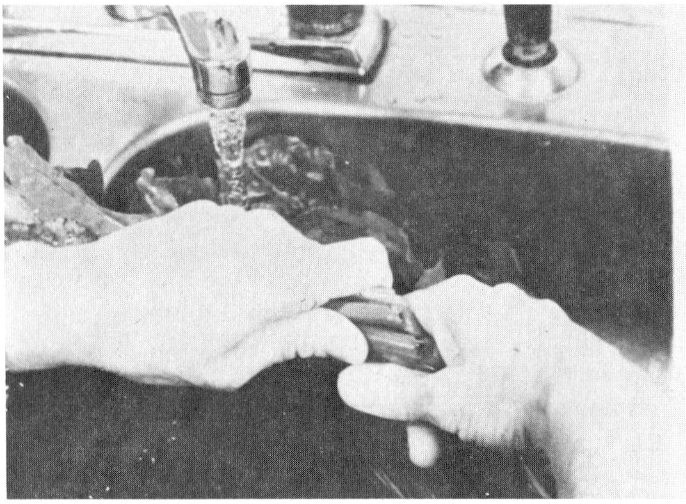

*Figure 8.16. About an inch of tops is left on each of the raw beets. The surface of the beet is not cut because the water soluble anthocyanin pigment in beets will be lost very rapidly into the cooking water, and the color of the beets will be poor.*

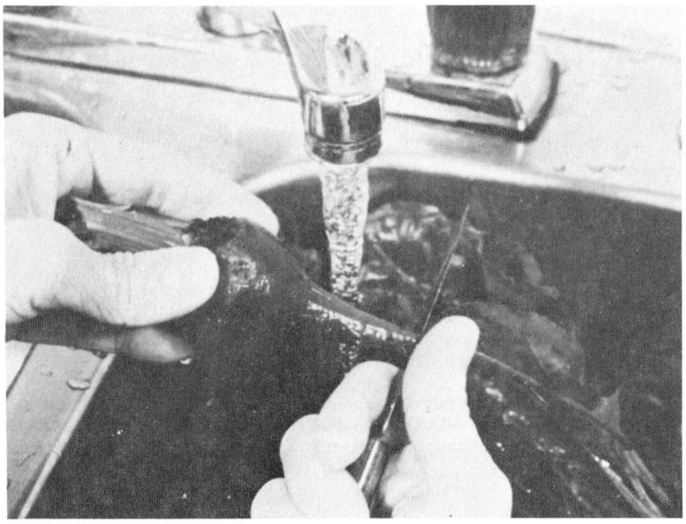

*Figure 8.17. About two inches of the root are left on. This also is a measure to minimize loss of color from the beets.*

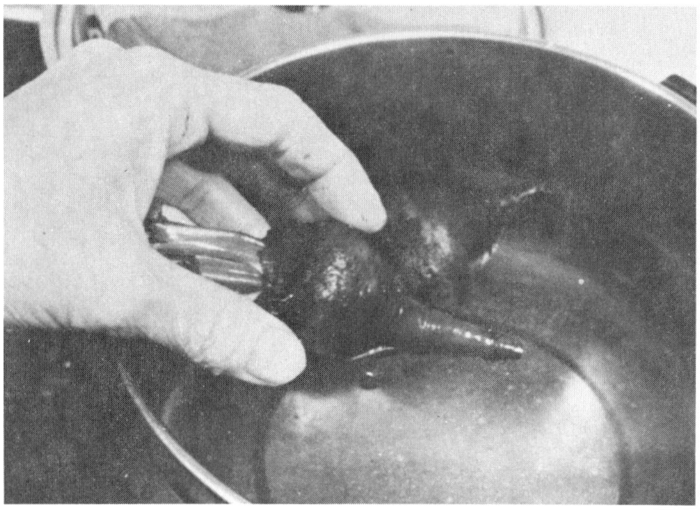

*Figure 8.18. The prepared, raw beets now are ready for boiling to tenderize them appropriately for making pickles.*

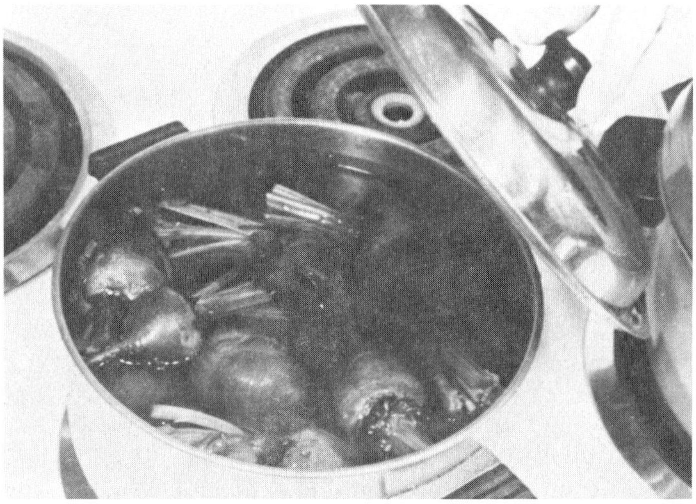

*Figure 8.19. A cover is used when boiling beets because they are mild in flavor, and their color will be enhanced by the use of a lid to trap volatile organic acids. The use of a cover also saves energy.*

*Figure 8.20. Beets can be pickled using ground spices. If preferred, whole cloves and stick cinnamon can be used. These are tied in a square of cheesecloth and simmered for 20 minutes in the syrup prior to the time the syrup is poured over the sliced beets in the jars.*

*Figure 8.21. Pickling cucumbers approximately three inches long and rather firm are better suited to the preparation of dill pickles than is the standard salad cucumber (shown in the lower right of picture).*

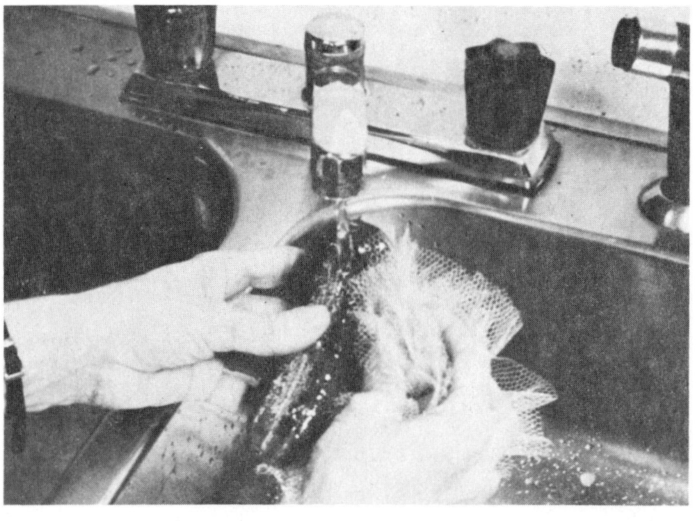

*Figure 8.22. A thorough scrubbing and careful removal of blossoms or other material are important steps in preparing cucumbers for making dill pickles.*

*Figure 8.23. The pickling salt, vinegar, spices, and water are combined in preparation for brining the cucumbers.*

## Dill Pickles

8 pounds cucumbers (about 3 inches long)
2 tablespoons pickling spices
2 or 3 small red chiles
8 cloves garlic (peeled and cut in half)
½ cup coarse Kosher salt
1 bunch dill plant (fresh or dried)
1 cup vinegar
Water (to cover cucumbers in crock)

1. Scrub cucumbers very well, being sure there are no blossoms or other foreign matter clinging to the cucumbers.
2. After draining well, place cucumbers in a 1-gallon crock. Mix the remaining ingredients (except the water) and pour over the cucumbers. Add water until the cucumbers are covered.
3. Stir solution well before placing a plate weighted with a glass of water over the pickles (or use a plastic bag filled with water) to hold the cucumbers below the surface and to minimize evaporation from the crock.
4. Place the crock in a cool, dark room where the odor of fermentation will not be objectionable. Check the crock daily and remove any scum which forms.
5. Allow to ferment for three weeks (or slightly less, if preferred) before packing the pickles in quart jars.
6. Strain the pickling brine and then pour into the filled jars, leaving a headspace of ½ inch.
7. Add a few sprigs of dill; close jars and process in water bath canner. Boil water actively for 15 minutes to process the pickles. Cool, check seal, label, and store as described for canning.

## Pickled Fish

1½ pounds firm white fish
¾ cup white vinegar (to cover fish)
⅓ cup Kosher salt
½ cup sugar
½ cup sherry or white wine
1 cup white (champagne) vinegar
1 tablespoon mixed pickling spices
1 tablespoon whole allspice
1 medium onion, thinly sliced and in separate rings

1. Cut fish into bite-size chunks and place in a quart jar.
2. Add ⅓ cup Kosher salt and vinegar to cover. Shake jar to dissolve salt.
3. Refrigerate, covered, for 5 to 7 days.
4. Drain off the brine and soak fish in fresh water for 2 hours before draining and rinsing fish.
5. Combine remaining ingredients in a quart jar and add the drained fish.
6. Cover jar and refrigerate for several hours.
7. For optimum quality, fish pickles are best when stored in the refrigerator for less than a month. For longer storage, fish pickles should be packed in pint jars following the initial few hours of refrigerated storage in the marinade and then processed in the pressure canner for 15 minutes at 10 pounds pressure, following the procedures for canning.

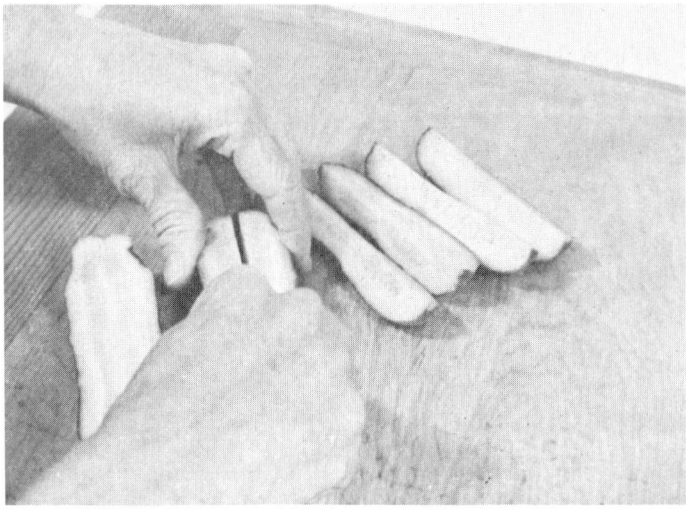

*Figure 8.24. After the fermentation period, the cucumbers are removed from the brine in preparation for filling the jars. If desired, the cucumbers may be cut into halves or quarters, rather than being left whole. Quarters can be packed more efficiently than the whole cucumbers.*

*Figure 8.25. By arranging strips of pickles vertically in the jar, the pickles can be packed efficiently and the convection currents in the jar during processing will insure excellent heating of the pickles in the water bath canner.*

*Figure 8.26. The brine used during fermentation of the pickles is strained to improve its clarity before being poured over the pickles in the jars. A half inch of headspace is left in each jar.*

*Figure 8.27. Sprigs of dill are added to the jars if they are available. In their absence, dill seeds can be added to enhance the flavor and appearance of the jars of pickles.*

*Figure 8.28. Dill pickles are processed in the water bath canner for 15 minutes in actively boiling water. Then they are removed and cooled on a rack away from drafts.*

*Figure 8.29. Firm, white fish is cut into bite-size pieces in preparation for pickling.*

*Figure 8.30. Pickling salt is poured over the fish in the jar to prepare for the brining of the fish.*

*Figure 8.31. The vinegar is added to the fish and salt.*

*Figure 8.32. After closing the jar, the jar is shaken thoroughly to dissolve the salt in the vinegar. Then the jar is refrigerated for 5 to 7 days to develop the flavor prior to heat processing. Note that this fermentation is not done at room temperature, as was true in making dill pickles. Refrigeration of fish is needed to prevent the growth of harmful microorganisms.*

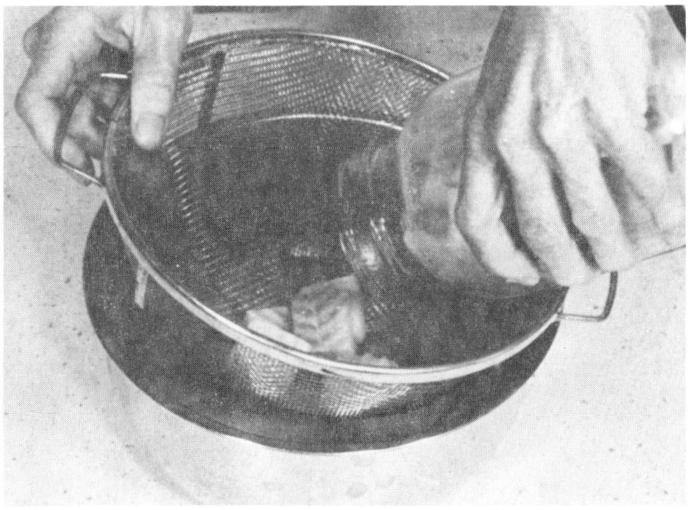

*Figure 8.33. After the brining period, the brine is drained from the fish and discarded.*

*Figure 8.34. A two-hour soaking period in fresh water helps to remove the saltiness from the fish, and a final spray rinse completes preparation for packing into the jar.*

*Figure 8.35. The remaining cup of vinegar, the sugar, wine, and spices are placed in the jar, and then the fish is added. For unrefrigerated storage or long-term storage, the jars are pressure processed for 15 minutes in a pressure canner at 10 pounds pressure.*

## Sauerkraut

17 pounds cabbage
½ cup Kosher or canning salt (iodine-free)

1. Wash cabbage well and drain. Shred cabbage after removing outer leaves and core.
2. Weigh 5 pounds of cabbage, transfer to a large bowl, and mix in thoroughly 3 tablespoons salt.
3. Transfer the cabbage-salt mixture to a large crock. Repeat the process described in Step 2 until the cabbage has all been salted and placed in the crock. Press down with wooden spoon until juice comes to the surface after each addition of cabbage.
4. Cover the cabbage with a clean piece of muslin, tucking the fabric down along the edges of the container. Cover with a plate which just fits inside the crock and which is weighted with a glass of water (or cover with a plastic bag filled with water). Be sure the cabbage is held beneath the surface.
5. Allow to ferment at room temperature (about 70°F) until fermentation stops and no more gas bubbles form, a storage period of approximately 5 weeks.
6. Heat fermented cabbage to simmering, being careful not to boil.
7. Pack hot sauerkraut into canning jars. Close and proceed as with other forms of canning in the water bath canner. Process in boiling water for 15 minutes for pints and for 20 minutes for quarts.
8. Cool, check seal, label, and store as for other canned goods.

# FRUIT PICKLES

## Peach Pickles

5 pounds (20 to 24) small peaches
Whole cloves
2 cups brown sugar
2 cups white sugar
1½ cups cider vinegar
½ teaspoon salt
1½ teaspoons powdered ginger
1 large piece of stick cinnamon

1. Dip washed peaches in boiling water long enough to loosen the skin. Peel, but do not pit.
2. Prepare syrup of the sugars, vinegar, salt, and ginger.
3. Place 2 whole cloves in each peach before placing the fruit in the syrup. Simmer fruit until fork tender. Let stand in covered kettle 8 to 12 hours.
4. Pack into hot, clean canning jars and cover with syrup. Add an inch of stick cinnamon and finish adding syrup to leave a headspace of ½ inch.
5. Close jars and process for 20 minutes in boiling water bath before cooling, checking seal, labeling, and storing.

## Watermelon Pickles

Rind of 1 large watermelon
¾ cup salt
6 cups sugar
1 quart white vinegar
7 teaspoons whole cloves
6 sticks cinnamon
1 piece fresh ginger root (optional)
½ lemon, thinly sliced

1. Remove all green skin and pink flesh of melon from the rind and cut rind into 1-inch pieces.
2. Cover rind with brine made by dissolving ¾ cup salt in 2 quarts of water.
3. After soaking in brine overnight, drain and rinse rind, cover with fresh water, and simmer rind until almost tender.
4. Tie spices in a rectangle of cheesecloth before simmering with the vinegar and sugar.
5. After simmering syrup for 10 minutes, add the drained rind; continue cooking until rind becomes translucent. Add boiling water to syrup if syrup becomes too viscous while the rind is being cooked.
6. Remove spice bag. Pack rind in canning jars and add syrup, leaving ½ inch headspace.
7. Close jars and process in boiling water in water bath canner for 10 minutes.
8. Cool, check seals, label, and store according to recommended canning procedures.

Yield: 6 pints

## Spiced Crab Apples

6 pounds crab apples with stems
1 quart cider vinegar
1 quart water
2 tablespoons whole cloves
2 tablespoons whole allspice
2 sticks cinnamon
1 cup cinnamon red hots
5 cups sugar

Yield: 7 pints

1. Make a small hole in the skin of each washed apple with a cooking fork (to reduce rupturing of skin during cooking).
2. Tie spices in a square of cheesecloth and heat with the other ingredients until the apples are almost tender.
3. Let apples stand in a covered container in the syrup overnight.
4. Remove spice bag; pack apples in canning jars and pour in syrup, leaving ½ inch headspace.
5. Close jars and process in boiling water in water bath canner for 15 minutes before cooling, checking for seal, labeling, and storing.

## Pickled Apricots

Pickled apricots may be prepared as described above. Substitute apricots for the peaches, but do not peel the apricots.

## OTHER RECIPES

### Antique Fruit

Seasonal fresh fruits
1 pint brandy
Sugar

1. Pour brandy in bottom of crock.
2. Add washed, fresh (or canned) fruits cut in small pieces.
3. Add a measure of sugar equal to the volume of fruit added. Stir well to mix fruit, sugar, and brandy.
4. Cover with a lid which rests on top of the crock, but which does not fasten securely. Let stand in a cool place for at least 6 weeks. Stir fruit once a week.
5. If desired, fruit can be added to the crock at any time in the same manner as described in steps 2 and 3 above. A variety of fruits makes a more interesting product.
6. If fruit is to be kept after it has achieved its desired strength of flavor, pack in canning jars with ½ inch headspace. Process for 10 minutes in boiling water in water bath canner. Cool, check for seal, label, and store as for other canned foods.

*Note:* Antique fruit can be served over ice creams, cakes, baked custard, or puddings.

*Figure 8.36. Whole oranges to be used for Rosy Spiced Orange Slices are simmered in water for an hour.*

*Figure 8.37. The drained, cooked oranges are sliced in wedges.*

### Rosy Spiced Orange Slices

10 medium thin-skinned oranges
2 cups water
2½ cups sugar
1 cup cinnamon candies (red hots)
1 teaspoon red food coloring

1. Place the whole oranges in a saucepan and cover with water. Simmer, covered, for 1 hour.
2. Drain the fruit; cut each orange into four to eight wedges.
3. In a saucepan, combine the sugar, water, cinnamon candies, and food coloring. Heat to boiling.
4. Arrange the orange wedges in a shallow baking dish in a single layer.
5. Pour the syrup over the fruit.
6. Bake uncovered for one hour at 350°F.
7. Transfer fruit to glass canning jars and pour syrup over the fruit.
8. Process in a water bath canner for 10 minutes in boiling water. (For storage of shorter duration, perhaps as long as six weeks, the orange slices may be stored in covered containers in the refrigerator without heat processing.)
9. Use the orange wedges as an accompaniment for meat, poultry, and fish.

*Figure 8.38. A syrup of sugar, water, and cinnamon candies, and red food coloring is prepared.*

*Figure 8.39. The orange wedges, arranged in a single layer in a shallow baking dish, are covered with the hot syrup.*

*Figure 8.40. If the wedges are to be used within six weeks, they may be placed in a plastic container and stored in the refrigerator. Longer storage requires canning in glass jars, with heat processing being done for 10 minutes in boiling water in a water bath canner.*

## Pickled Eggs

**Red Beet Eggs**

1 dozen hard-cooked eggs, peeled
1 cup red beet juice
½ cup red wine vinegar
½ cup rose wine
1 teaspoon brown sugar
2 teaspoons seasoned salt
Few slices of beet pickles

1. Place the peeled eggs in a quart jar.
2. Combine the other ingredients in a saucepan and heat to boiling.
3. Pour the hot pickling solution over the eggs; close lid tightly.
4. Refrigerate for at least a week for small eggs and two weeks for larger ones. Eggs may be stored for several months prior to use, but they must be kept refrigerated since the jars are not heat processed.

**Dilled Eggs**

1 dozen hard-cooked eggs, peeled
1½ cups white vinegar
1 cup water
¾ teaspoon dill seed
¼ teaspoon white pepper
1 tablespoon salt
¼ teaspoon mustard seed
½ teaspoon onion juice
½ teaspoon minced garlic

1. Follow the procedure outlined for making Red Beet Eggs. Refrigerated storage also is required for the Dilled Eggs.

**Dark and Spicy Eggs**

1 dozen hard-cooked eggs, peeled
1½ cups cider vinegar
½ cup water
1 tablespoon dark brown sugar
2 teaspoons granulated sugar
1 teaspoon mixed pickling spice
¼ teaspoon liquid smoke or hickory smoke salt
2 teaspoons salt

1. Follow the procedure outlined for making Red Beet Eggs. Refrigerated storage also is required for the Dark and Spicy Eggs.

*Note:* Pickled eggs are useful as a nourishing snack, as hors d'oeuvres, garnishes for salads, and deviled eggs.

**Dijon-Style Mustard**

2 cups dry white wine
1 large onion, chopped
3 cloves garlic, pressed
1 4-ounce can dry mustard
3 tablespoons honey
1 tablespoon vegetable oil
2 teaspoons salt
Pepper seasoning

1. Combine wine, onion, and garlic in saucepan.
2. Heat to boiling, and simmer five minutes.
3. Cool before straining wine mixture into dry mustard, being sure to stir constantly to achieve a smooth slurry. Discard the onion and garlic.
4. Blend in the remaining ingredients.
5. In a saucepan, heat slowly and stir constantly until thickened.
6. Cool before storing in a covered container in the refrigerator. This product may be kept indefinitely in the refrigerator; heat processing is not required for refrigerator storage.

*Note:* This mustard is not a pickle, but it does serve as a flavorful relish and is well suited to use with various cucumber pickles.

*Figure 8.41. After the wine, onion, and garlic in the recipe for Dijon-Style Mustard have been simmered together for five minutes, they are allowed to cool before being strained through cheesecloth.*

*Figure 8.42. Thorough blending of the liquid with the dry mustard is required to achieve a creamy, smooth product.*

## Homemade Chili Sauce

15 to 20 ripe tomatoes (3 quarts, chopped)
6 medium onions
6 green and red sweet peppers
1 3-inch stick of cinnamon
1 teaspoon whole cloves
1 teaspoon whole allspice
1 pint vinegar (5% acetic acid)
1 tablespoon salt
1 cup sugar
1 cup brown sugar

1. Blanch the tomatoes in boiling water for a few seconds to loosen the skins. Peel and chop the tomatoes.
2. Peel and chop the onions.
3. Remove the stem and seeds from the peppers and chop them.
4. Tie the cinnamon, cloves, and allspice in a square of cheesecloth.
5. In a covered kettle, simmer the tomatoes, onions, peppers, and spices for 40 minutes. Remove the spice bag.
6. Add the vinegar, salt, and sugar to the tomato mixture. Heat to boiling and boil actively for 5 minutes.
7. Immediately ladle the chili sauce into clean, hot pint jars, leaving ½ inch headspace. Seal with canning lids prepared as described above.
8. Place in a water bath canner and process for 15 minutes when the water boils in the canner. If preferred, pressure canning for 10 minutes at 5 pounds pressure can be used.

*Note:* Homemade chili sauces are a more earthy, warm tone than commercial chili sauces.

## Chili Sauce ala Padgett

1 gallon (6¼ lb.) ripe tomatoes (peeled and cored before measuring)
⅔ cup chopped white onion
1½ cups white sugar
1 teaspoon grated nutmeg
¾ teaspoon Tabasco sauce
½ teaspoon curry powder
2 cups vinegar
5 teaspoons salt
2 teaspoons ginger
1 teaspoon cinnamon
1 teaspoon mustard
1 teaspoon dry mustard

1. Put onions and tomatoes through food chopper.
2. Add all other ingredients and simmer 3 hours until thick, stirring frequently to prevent burning.
3. Boil until sauce is the desired consistency.
4. Ladle into pint jars.
5. Seal and transfer to water bath canner.
6. Process for 15 minutes in boiling water in the water bath canner.

*Note:* If desired, ¾ teaspoon red pepper may be substituted for the Tabasco sauce.

PEACH NUT
CONSERVE

# Chapter 9.
# Preserving with Sugar

## THEORETICAL FOUNDATIONS

### Preservation Principles

Jams, jellies, and related products are representative of foods which derive their prolonged shelf life by virtue of a high sugar content. The preparation of these products requires a rather high proportion of sugar in the recipe itself and further concentration of the sugar by evaporating water from the boiling fruit mixture. The necessary concentration of sugar for safe preserving action is achieved by heating the mixture to a temperature 8 to 9°F. above the boiling temperature of water.

The preserving action of sugar is achieved because of the effect sugar has on osmotic pressure. In jams and jellies, the sugar is in solution while these foods are stored on the shelf. This sugar solution contains a concentration of sugar significantly greater than the concentration of soluble materials contained within the cells of microorganisms. The cell walls serve as semipermeable membranes or barriers between the medium of the jelly and the internal cellular medium of the microorganism. As a result of the semipermeability of the cellular membrane, there is a difference in the concentrations of the solute on either side of the membrane. This situation is true because sugar and other types of selected materials, such as mineral salts, cannot pass through the membrane. However, water can pass through. It is the ability of water to pass through the semipermeable cellular membranes of microorganisms which is the key to this preserving action of sugar.

The boiling of jams and jellies to achieve the necessary sugar concentration does, of course, have a hygienic effect on the product by destroying microorganisms which may be present at the time of

*Figure 9.1. The large amount of sugar used in making jams, jellies, and preserves acts as a preservative.*

*Figure 9.2. The concentration of sugar needed to achieve the desired consistency and preserving action is reached at a temperature 8 to 9°F. above the boiling point of water.*

preparation. However, there is potential for recontamination following this heating period. The problem of shelf storage then becomes contingent upon the provisions for eliminating these subsequent infections of the product by microorganisms. It is in this context that sugar plays its preserving role.

If microorganisms are allowed to enter the jam at any time after cooking, the microorganisms are in a hostile environment for survival and reproduction. This is due to the fact that the high sugar concentration outside the microorganisms will cause osmotic pressure to develop in which the balance of pressure draws water from within the cells of the microorganisms into the surrounding medium, i.e., the jam or jelly. Because of this situation, the microorganisms will be dehyrated and die.

The hostile effect of sugar is enhanced by the acidity of the environment, too. As has been noted previously, many microorganisms do not reproduce and flourish in a distinctly acidic environment. The pH of jams and jellies with good gel strength is 3.3 or even lower, clearly an acidic range contrary to the needs of microorganisms.

## Theory of Gel Formation

### Definition of a gel

A gel is defined as a colloidal system which has a degree of rigidity sufficient to keep it from flowing. In a gel, the solid constitutes the continuous phase and the liquid the discontinuous phase. In other words, there is some solid substance which makes a continuous network throughout the system. This network traps or hydrogen binds the liquid in many discrete, non-continuous regions between the mesh of the continuous phase. Specifically, in pectin gels, pectin constitutes the continuous, solid phase of the system. Fruit juice represents the liquid portion of the gel, thus being designated as the discontinuous phase.

*Figure 9.3. Low acidity of the product (pH of 3.3 or lower) aids in preventing growth of microorganisms. Lemon juice often is used to increase acidity of jams, jellies, and preserves.*

### Characteristics of a gel

A gel has physical properties which are influenced by the conditions utilized in preparing the product as well as by the ingredients it contains. One aspect of gels of concern from a food quality vantage point is the consistency of the gel. Variations in consistency may range from very tender gels which are deformed readily by only minimal pressure to very firm, rigid or rubbery gels. The consistency of pectin gels is important to their perceived quality since it is anticipated that jams and jellies can be spread relatively easily, yet not actually flow. This aspect of gel quality is evaluated by accepted methods of measuring tenderness, as is done with the penetrometer.

The consistency of a gel is not a totally static state. The continuous network of a gel undergoes changes, notably a tightening, during storage. Thus, assessment of consistency needs to be conducted after a gel has been aged at least six hours at the temperature planned for testing.

Another characteristic of gels is that they may exhibit syneresis (drainage of some of the liquid or discontinuous phase from the solid, continuous phase). The degree of syneresis is influenced by the type of pectin used in forming the gel. Assessment of the problems of syneresis is made after a gel has been sliced or cut and allowed to drain for a designated period of time. The cutting of a gel cleaves numerous areas where liquid has been trapped by the continuous phase, thus releasing some of the discontinuous (liquid) phase. Gels which are not disrupted by such cutting ordinarily exhibit little, if any syneresis.

### Structure of a gel

A gel is a random structure in which molecules of the material comprising the continuous phase are cross-linked to form a network. The cross-linkages ordinarily are hydrogen bonds between molecules. These secondary bonds can be established when portions of molecules containing oxygen and hydrogen are in close proximity to each other at cool temperatures. The cool environment is necessary if the hydrogen bonds are to have some degree of permanency. If the product is quite warm, the hydrogen bonds will be only transitory because of the active movement of the molecules within the medium. Without relatively permanent hydrogen bonding, the system will remain a sol rather than becoming a gel. Numerous hydrogen bonds with a high degree of stability are required to make a network of molecules sufficiently rigid for the gel to be established.

The continuous network of a gel often is described as being like a brush heap because there is apparently almost no systematic organization of structure. There may be areas of a gel where there are many hydrogen bonds between molecules and where only small spaces are surrounded by the continuous network. There may be other places where there is a more widely spaced arrangement. The frequency of hydrogen bonding and concentration of molecules of the continuous network significantly influence the strength or rigidity of the gel.

This brush heap of continuous-phase molecules surrounds or traps the molecules of the liquid phase of the system. Interestingly, it is possible for the continuous phase to hydrogen bond not only with itself, but also with the liquid phase. When hydrogen bonds are established between the solid and liquid phase molecules, the liquid molecules are immobilized and locked securely into the total framework of the gel. This binding of liquid along the surface of the molecules comprising the network further restricts the space available between the various parts of the structural network. However, there will be space available between the molecules of the network. These pockets in the network thus entrap free water, i.e., water which has not been hydrogen-bonded to become, in a sense, a part of the actual structure of the gel.

When a rather compact network of the solid has been established and further strengthened by the hydrogen bonding of some of the liquid, a gel forms. Although there will be some continuing formation and disruption of the hydrogen bonds within the gel, the structure appears visually to be established firmly. Some of this structure is completely immobilized by hydrogen bonds, as is some of the liquid. However, there will be the potential for unbound liquid to drain from the gel structure if its physical barriers (the continuous phase) are modified by cutting through the gel.

## PECTIN GELS

### Pectic Substances

The pectic substances are a family of chemically-related compounds which aid in holding the cells of fruits together. These compounds are classified as carbohydrates because they are derivatives of a simple carbohydrate, galactose. This base makes pectic substances unique among the polysaccharides because glucose is the usual building block for the polysaccharides occurring in foods. To add still another dimension to this uniqueness, the basic structural unit of pectic substances is not galactose itself, but instead is the uronic acid derivative, galacturonic acid. This structure has an organic acid radical on the carbon which is external to the ring structure, and it is this acid radical which is of significance in studying the chemical aspects of pectic substances and their ability to facilitate gel formation. The chemistry illustrating pectin is presented in the Appendix.

The pectic substances undergo chemical changes as the fruit traverses through its normal cycle from immature to ripe to overripe. These changes involve modification of functional groups and molecular size. They may be observed and traced by noting the physical changes related to texture of the fruit. The transition from very hard to somewhat yielding to very soft represents the textural range which accompanies the chemical modifications in the structures of the pectic substances.

#### Protopectin

In immature fruit, the pectic substance is protopectin. This compound is a linearly shaped molecule composed of hundreds of units of galacturonic acid linked by a 1,4-alpha linkage. This molecule does not have the ability to form a gel. The presence of pectic substances in the form of protopectin is identified by the very firm nature of the fruit in which it occurs. From the standpoint of gel preparation, this pectic substance is of little assistance or interest. However, this parent compound can be converted by enzymes in the maturing fruit into a more useful form. Also, some small amount of conversion of protopectin to pectin may occur as the result of a prolonged cooking period in jelly making.

#### Pectin

The term *pectin* designates a group of substances capable of forming a gel when present in the correct concentration. Included in this general term are both pectinic acids and salts of pectinic acids, known as pectinates. Pectin indicates any of the pectic substances which can establish the continuous network needed for gel formation and does not differentiate between the forms of the compound. Pectin formation occurs as the result of enzymatic action on protopectin. The enzymes effective in this transformation are polygalacturonase and pectin esterases. Warm temperatures or bruising of fruit hastens the chemical transformation to this type of pectic substance. Accompanying this chemical change is the gradual softening of the structure of the fruit. The accelerated softening of bruised or damaged fruit, compared with the normal structure, attests to the effectiveness of the enzymes in the fruit. The softening of fruit held at room temperature in comparison with refrigerated fruit also serves as testimony to the conditions influencing the activity of enzymes.

The chemical transition between protopectin and pectin involves esterification (specifically methylation) on the carbon external to the ring, i.e., the carbon in the acid radical of galacturonic acid. Ester formation occurs between the galacturonic acid radical and a molecule of methanol, yielding a methyl ester. This methylation process appears to be the vital change needed to enable gel formation to occur. Not all of the entire chain of galacturonic acid units will be esterified, but many will be.

Differences between pectins from different fruits have been noted. Although all pectins apparently are very large molecules, there is considerable difference between various sources. The two commercially important sources of pectin are apples and oranges. The molecular weight of the pectin from apples is approximately 280,000, but that found in the rind of oranges is only about 220,000. Yet despite this relatively great difference in chain length, both of these pectins have outstanding gel-forming properties.

Within the general classification of pectin a distinction is made chemically on the basis of the presence of a salt on some of the acid radicals or its absence. An example of a salt would be the addition of calcium on the acid radical of the sixth carbon. Compounds having these salts on some of the galacturonic acid units are classified as pectinates. Those without the salt formation are termed pectinic acid. Either of these forms has the ability to form gels. Hence, there is not a need to make a distinction between them.

### Pectic acid

As fruit continues to mature, the pectin contained in the barely mature fruit will begin to be transformed into pectic acid. Pectic acid no longer possesses the ability to form gels. Consequently, the practical application of this transition is to select fruit for jam and jellies which is barely ripe. This can be selected by visual observation of color and by noting the firmness of the fruit. When fruit yields very readily to slight pressure, the pectic substances have progressed beyond the pectin stage to the formation of pectic acid. Because of its loss of gel-forming ability, pectic acid is not of practical use. Therefore, overripe fruits are a poor choice in making pectin gels.

The reason for the loss of gel-forming ability in pectic acid is not known absolutely, but it appears that it may be influenced significantly by the loss of methyl groups from the structures. There is considerable demethylation which takes place as the fruit progresses to the overripe stage. There also may be some reduction in the molecular size of the molecules by hydrolysis of some of the linkages between the galacturonic acid units. The shortened chains which result would be more soluble and thus less capable of forming the necessary hydrogen-bonded network. The very long chains are inhibited in their movement in the system by their physical bulk, but the shorter chains effected by hydrolysis are more mobile.

## Definitions

The various pectin-gelled products made from fruits are designated by different names to differentiate between the parts of the fruits and the use of additional ingredients.

### Jams

Jams are made utilizing relatively small pieces cut from the fruits. The pieces of fruit are recognizable and are valued for their contribution to the texture of the finished product.

### Preserves

Preserves are very similar to jams; in fact, their identity often tends to merge with jams. The official distinction between these two products is that preserves are made utilizing larger pieces of fruit than are included in the preparation of jams. In fact, some preserves are made with whole fruits, as often is the case with raspberries and other berries.

### Conserves

An interesting and pleasing variation from preserves is made by adding chopped nuts to preserves. This addition adds considerable interest to the texture, as well as providing a subtle flavor overtone. These products are called conserves.

### Marmalades

Marmalade actually is simply an example of preserves in which thin slices of the citrus rind are included. Tradition has caused citrus preserves to remain identified by the unique title of marmalade.

### Butters

Fruit butters represent yet another way of preserving fruits for use as spreads. In these products pureed fruit replaces the use of pieces of the fruit itself. The level of sugar in fruit butters is significantly less than is found in the other pectin-containing products. For this reason, the sugar is not able to control microorganisms, and the butters will have limited storage life unless canned and heat processed in boiling water in a water bath canner. The butters ordinarily contain a rather large amount of spices, but the level is not sufficient to preserve the product without heat processing.

*Figure 9.4. Apple butter is made with too low a sugar content to prevent growth of microorganisms. Consequently, jars must be closed and processed by water bath canning to sanitize and seal the product.*

### Jellies

Jellies are judged by many people to be the most beautiful of the pectin-containing products. The sparkling and transparent appearance is enhanced by the delicacy of color associated with well prepared jellies. Added to this appeal is the smooth and slightly firm, yet yielding texture of jelly. Jellies gain their distinctive nature because they are prepared only from the extracted juice. No pulp is contained to mar the clarity of the product. The color remains bright when fundamental principles of preserving are observed.

## Functions of Ingredients

### Pectin

Pectin serves as the main structural component of the network or continuous phase of the gel structure. For a successful gelled product, it is essential to have a high enough concentration of pectin to permit a structurally sound random framework to be formed from the hot sol. The effectiveness of the pectin in forming a gel is influenced by the amount of methylation in the pectin molecules. This level is determined by the maturity of the fruit, with barely ripe fruit having a much higher level than either immature or overripe fruit.

### Sugar

Sugar in pectin gels serves other important functions as well as that of preserving, which was discussed previously. One important role of sugar is related to its hygroscopic nature. Sugar attracts water very readily from its environment as a result of the numerous hydroxyl (–OH) groups on its surface. Sugar in pectin gels attracts water which might otherwise be attracted to the surface of pectin. By aiding in eliminating the layering of water on the pectin molecules, sugar helps to make it easier for pectin molecules to come in direct contact with each other, thus facilitating hydrogen bonding to form the necessary structural network of the gel.

This same hygroscopic quality helps to hold water within the loose framework of the gel. This action is useful not only in establishing a gel rather than a sol, but it also aids in minimizing syneresis.

The flavor contribution of sugar certainly should not be ignored. A pectin gel could be beautiful to behold and easy to spread, yet it would lose much of its appeal if it lacked the sweet taste afforded by the sugar in the product.

To an extent, sugar also influences the color of the finished product. If a large amount of sugar is included in the recipe, only a very short period of boiling will be necessary to achieve the necessary concentration of sugar. However, if only a small amount of sugar is included, it will be necessary to cook the product a fairly long time before the necessary concentration of sugar is achieved. During this extended cooking period, the color of the product darkens from its original, relatively bright color to a somber shade. This change is influenced greatly by the fact that some of the sugar breaks down chemically during cooking of the mixture. First hydrolysis occurs, followed by some breakdown in the ring structure itself. The mixture of organic acids, aldehydes, and ketones resulting from the extended boiling period will make the finished product less attractive in color.

### Fruit and fruit juices

The fruits and fruit juices contribute some pectin; its role has been described above. However, the fruits and their juices provide important sources of the acids needed for successful gelation. The pH of the fruit mixture is important in determining the ability of the product to gel. A pH of 3.3 is ideal for making pectin gels. At this level of acidity, the pectin molecules have a minimum electrical charge on their surface. This reduction in charge makes it much easier for pectin molecules to approach each other and hydrogen bond. At either a higher or lower pH the electrical charges on the surface of the pectin will begin to build again, and gel structure cannot be established so readily. The acidity of the fruit will vary somewhat, depending on the maturity. However, the barely ripe fruit is still quite high in acid content and has the additional advantage of containing pectin in far greater concentration than would be available in the very green fruits.

The acid in pectin gels obviously plays a very key role in the preparation of pectin gels. In addition to this essential contribution, acids add to the intriguing flavors of pectin gels.

## Selection of Ingredients

### Pectin

When some fruits are barely ripe, they contain enough pectin to enable them to gel without using added pectin. Apples and citrus peels are familiar examples of fruits rich in pectin. In contrast, such favorites as strawberries, raspberries, peaches, and apricots are not high enough in pectin to make a satisfactory gel with a reasonable period of cooking.

To make satisfactory gels using fruits naturally low in pectin, it is practical to use a commercial pectin to supplement the natural content. These commercial pectins are available in two forms: liquid and powdered. The sources utilized for preparation of either of these types of pectin products are the skins and cores of apples and the albedo (white portion) of citrus fruits. The latter source often is used since the production of frozen citrus juice concentrates generates huge quantities of the rind, which is used more suitably for the production of a salable product (pectin) rather than being discarded simply as waste.

Powdered pectin is a useful product for persons who make small quantities of jams or jellies at infrequent intervals. The merit of powdered pectin is its longer shelf life in comparison with an opened and partially-used bottle of the liquid pectin. It is possible to use only the amount of the powdered pectin required for the current project and then to seal the remainder of the package for later use. The powdered pectin requires rehydration, a process which simply requires that the pectin be added at an earlier point in the cooking process than is done when liquid pectin is used. For successful products, it is important to note this difference in procedure.

Liquid pectin is bottled in a fluid form which does not require hydration. This product is very convenient to use, but it undergoes a rather rapid breakdown in its chemical structure to pectic acid when the opened bottle is stored for later use. The change in the structure of liquid pectin can be noted visually. Liquid pectin is naturally very viscous, in fact is so viscous that it flows with some difficulty from the narrow neck of the bottle. However, pectin which has undergone extremely long shelf storage, has been improperly sealed, or has been opened and then stored for awhile may become noticeably less viscous. It may liquefy to the point where it can be poured like water from the bottle. Actually, this product is no longer pectin, but rather has been transformed into breakdown products which no longer possess the ability to form gels. If such a pectin is used in making jams or jellies, the end product will not gel properly. Simply as a sensible precaution, it is wise to note the viscosity of the pectin in the bottle before using it. There is no way to reverse this detrimental change in chemical structure. A new bottle of pectin must be used. Liquid pectin is added near the end of the boiling period in the preparation of jams and jellies to avoid breakdown of the pectin to pectic acid during the cooking period.

### Fruits

If the fruit is to be preserved without the addition of commercial pectin, a barely ripe fruit is the degree of maturity to select. This stage of ripeness has the advantage of being relatively acidic as well as being at the maximum level of pectin. However, the color of the fruit and the flavor may not be as pleasing at this stage as they will be later when the fruit matures a bit more. If commercial pectin is being used, it is possible to select fruits with the flavor characteristics and the color desired.

The size of fruits is not of great significance when jams and jellies are being prepared because the fruit will be cut into small pieces for jams and only the juice will be used in preparing jellies. Sometimes the smaller fruits are more flavorful than the very large ones, and the price usually is less at the market. If fruits for preserving are grown at home, the relative price difference does not apply, but the difference in flavor should be considered.

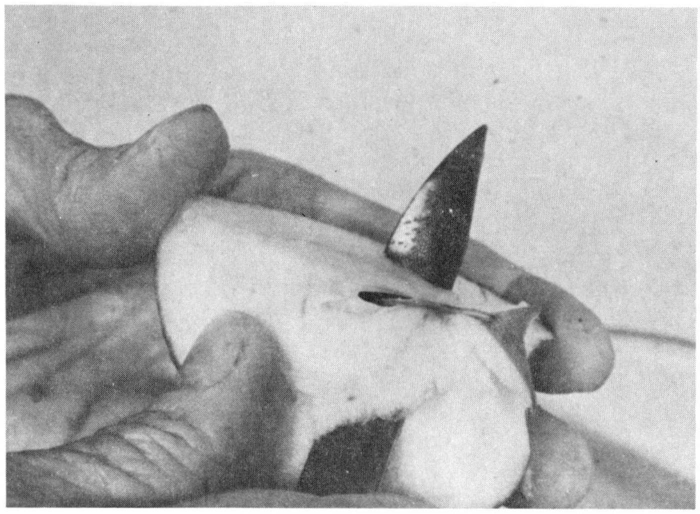

*Figure 9.5. The pectin content of apples is relatively high, making it possible to use them for jams and jellies without adding pectin. In fact, apples serve as a source of commercially-produced pectin.*

*Figure 9.6. Marmalades can be made at home without added pectin because the albedo (white) of the rind is an unusually rich source of pectin. The rinds of citrus which have been utilized for manufacturing juice concentrates are used for manufacturing pectin on a wide scale.*

Many different fruits are well suited to preserving, either as individual jams and jellies or as combinations of more than one fruit. Availability, cost, and personal preferences are important considerations when considering the preparation of these products in the home.

### Acids

Some fruits are naturally low in acid content; some may be low because they have matured and are beginning to be overripe. If the fruit itself is not at approximately pH 3.3, it is necessary to add an acid to modify the acidity of the mixture. Lemon juice often is added to these fruits to bring them to the pH range needed for jelly-making.

Fruits which may be expected to have a sufficient acidity without added acid include citrus fruits, grapes, tart apples, and berries. If pH paper

*Figure 9.7. Commercially-prepared pectin is available in the powdered form, which is convenient when it may be held on the shelf for a long period of time. Powdered pectin is added at the beginning of jelly making so that the pectin will be rehydrated completely before the end of the boiling period.*

*Figure 9.8. Liquid pectin is added near the end of the jelly making operation to avoid chemical breakdown to the non-gelling pectic substance, pectic acid.*

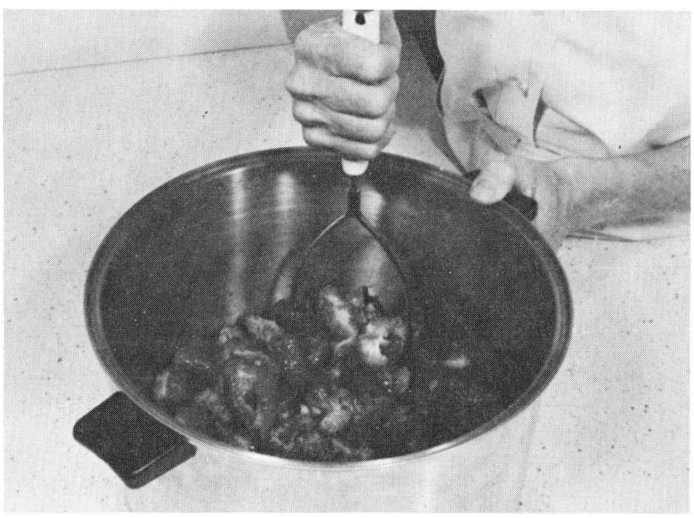

*Figure 9.9. The size of fruits for making jams and jellies is not important because they will either be cut into small pieces or squeezed to extract their juice. However, it is important to select just ripe fruits to obtain optimal acidity and pectin content.*

*Figure 9.10. Although lemon juice is the acid usually used to increase acidity of jellies, vinegar can be an effective acid in special products, such as in this green pepper jelly.*

is available, it is wise to check the acidity of fruits, even those which might be assumed to be sufficiently acidic. This assures that harmful variations from the recommended acidity will be noted and can be corrected so that the finished product will be of the highest possible quality. This check becomes particularly important when colorful, mature fruits have been selected.

**Sugar**

Granulated sugar (sucrose) is the form of sugar used traditionally in the preparation of jams and jellies. The amount of sugar to use is a matter of importance. Sometimes people decide to modify a recipe by reducing the amount of sugar from the quantity recommended. This is a dubious practice which has a detrimental effect on the quality of the finished product. As pointed out earlier, the addition of a limited amount of sugar in the recipe requires that a longer boiling period be used to achieve the necessary concentration of sugar. The effect of this longer boiling period is a smaller total yield of the jam or jelly. It also causes the product to be darker and less translucent because the longer cooking concentrates the pectin and caramelizes some of the sugar.

Honey may be used as a sweetener in jams and jellies, but its liquid content needs to be considered when it is replacing part of the sugar in the recipe. If the recipe contains powdered pectin or liquid pectin, up to a cup of honey can replace an equal quantity of granulated sugar when making approximately 6 jelly glasses full or up to two cups of honey can be used in a larger batch. When no pectin is added to a jelly recipe, a maximum of half the sugar in the recipe may be substituted by an equal amount of honey.

Another sweetener sometimes used is corn syrup. Again, the liquid content of the sweetener must be considered in making substitutions. Up to a fourth of the sugar can be replaced by an equal amount of

Table 9.1. Sugar substitutions in jams and jellies

| Sweetener | Amount of sweetener/cup of sugar | Limitations to substitution |
|---|---|---|
| Honey | 1 c./1 c. sugar | For 6 glasses of jelly if PECTIN is used |
| | 2 c./2 c. sugar | Larger batch than 6 glasses if PECTIN is used |
| Corn syrup | 1 c./1 c. sugar | Maximum of ¼ of the sugar if NO PECTIN added |
| | 1 c./1 c. sugar | Maximum of ½ of the sugar if POWDERED PECTIN used |
| | 1 c./1 c. | Maximum of 2 c. per batch if LIQUID PECTIN used |

corn syrup if the recipe does not contain added pectin. When the recipe contains added powdered pectin, the substitution can be increased to half the amount of sugar in the recipe; with liquid pectin recipes, a maximum replacement level is two cups of corn syrup for a comparable amount of granulated sugar.

It should be noted that the above substitutions result in a greater variation of sugars in the final product. There will be some inversion of sucrose to yield a mixture of sucrose, glucose, and fructose when granulated sugar is used. This is true regardless of whether cane or beet sugar is used because the sugar in both of these foods is in the form of sucrose. On the other hand, honey contains a mixture of sugars, with the sugar of primary interest being fructose. Corn syrup is also a mixture of compounds, with the predominant carbohydrates being maltose and glucose. The mixtures of sugars help to insure that sugar crystals will not form in the stored product. However, crystallization is not a significant problem in properly formulated and prepared jams and jellies.

Sometimes variations of jams and jellies are prepared in which the sugar content is significantly lower than the normal level or in which a sugar substitute is used in place of sugar. These products require freezer storage for long storage and refrigerated storage because they lack the preserving action of sugar solutions on bacteria and other microorganisms. The protective action of sugar is afforded when jams and jellies are cooked to a temperature 8 or 9°F. above the boiling temperature of water. This temperature corresponds to a sugar concentration of 65 percent.

*Figure 9.11. Freezer storage is utilized in special jams made with sugar levels too low to attain protection against growth of microorganisms at room temperature.*

## STEPS IN PREPARING PECTIN GELS

### Assembling Equipment

The most important single piece of equipment for making jams and jellies is a large kettle, preferably of heavy aluminum or other material which conducts heat well. The large size is essential because a full rolling boil is needed to concentrate the sugar by evaporating the water quickly. This aids in minimizing chemical change in the pectin and in the sugar. The heavy metal provides useful protection against possible scorching of the very hot, sugar-containing mixture.

If jellies are being made with fresh fruits, a muslin bag is needed to facilitate removal of the juice. Expression of liquid from the pulp is aided greatly by placing the fruit-filled muslin bag in a colander equipped with a support stand.

*Figure 9.12. A large kettle is required for making jams, jellies, and preserves so that rapid evaporation can be accomplished to increase the sugar concentration before the pectin breaks down.*

Other important items for preparing the gels include measuring cups and spoons (with at least one cup with a 1-quart capacity), paring and utility knives, a potato masher, wooden spoons with long handles, and a clock with a second hand.

Jelly glasses make convenient jars for these products, but canning jars also are suitable. Sturdy glasses also may be used. Any containers selected need to be able to withstand the high temperature of the product when it is poured into the glass directly from the range. They should be free of any imperfections and chips. If jars are to be sealed with canning closures, new lids should be available, although the bands may be reused. If paraffin is to be the sealing technique used, the paraffin should be placed in a relatively large can which has been washed and thoroughly dried before the paraffin is placed in it. This can, in turn, should fit into a pan at least an inch wider than the diameter of the can.

*Figure 9.13. Paraffin, because of its flammability, should be heated very carefully, preferably in a metal can resting in a pan containing some water.*

## Preparing the Containers

Careful preparation of the containers and closures is particularly important in making jams and jellies because of the use of a canning technique known as *open kettle canning.* This method requires the use of sterile equipment for optimum shelf life of the finished product. Furthermore, this technique also results in unusually hot syrup being poured into the containers. Prior heating of the glass containers is important to help in minimizing heat shock to the glass when the jams or jellies are ladled into the containers.

All containers and closures should be washed thoroughly in hot, soapy water and then rinsed very thoroughly in hot water, being careful to get all traces of detergent removed. The rinsed jars then are placed in a pan of water deep enough to cover all of the jars, and they are boiled for 10 minutes. These containers may be left in the pan of hot water with the heat turned off under the pan on the range or they may be moved into the oven (set at 200°F) or the pan may simply be moved out of the way on the kitchen counter. However, it is important that the jars be kept hot while the jam is being prepared.

After the closures have been washed in hot, soapy water and rinsed, they should be treated according to manufacturer's directions. Attention to the treatment recommended is important to maintaining the ability of the compound on the lid to achieve a seal. Improper treatment can result in detrimental changes to the sealing material, with the result that the jar will not seal.

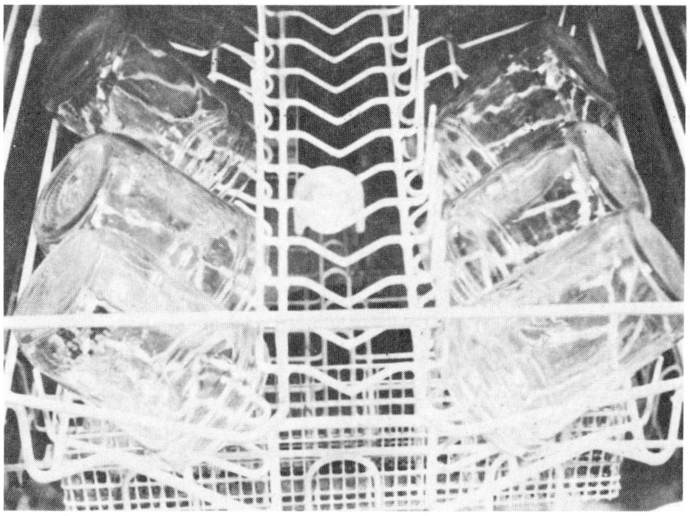

*Figure 9.14. Jars for jams, jellies, and preserves can be prepared to receive the very hot fruit products by heating them in a dishwasher.*

## Jellies

### Preparing the juice

Fruits to be used for making jellies need to be washed thoroughly, and any blemishes should be removed. Most fruits then require a brief cooking period with a small amount of water added to the fruit. The cooking is done to help soften the fruit so that the juice can be extracted more easily. Very juicy, rather soft fruits, such as berries can simply be crushed without heating.

After the fruit is softened sufficiently for the juice to be extracted, the fruit is transferred into the muslin jelly bag, which has been dampened with water. To obtain very clear juice, the fruit should be allowed to drain naturally through the bag. However, more juice can be extracted if the bag is twisted tightly, pressed, or squeezed. This severe treatment damages the cell structure significantly and results in a juice which is quite cloudy. The clarity of the juice can be restored, in part, by straining the extracted juice a second time, either through two layers of cheesecloth or a jelly bag. These straining devices should be dampened before the juice is placed in them. This facilitates recovery of a maximum amount of juice.

It is possible to make jellies of excellent quality using commercially-prepared fruit juices, too. The use of these products is very simple because the preceding steps of preparing the fruit and extracting the juice are eliminated.

**Use of pectin**

Some jelly recipes are based on using the pectin contained in the fruit as the only pectin for the gel structure. When pectin is not being added to the batch of jelly, it is wise to check the approximate amount of pectin naturally present in the fruit. The pectin content of the extracted juice can be checked conveniently by either of two methods. If a jelmeter is available, this is a convenient means of checking pectin content. The jelmeter is a simple device designed in a fashion very similar to that of a pipette. The device is marked to permit a rough assessment of the viscosity or rate of flow of the juice through the jelmeter. A very slow rate of flow is indicative of a high pectin content, and the juice will be appropriate for use in jelly making without added pectin. If the juice flows rapidly through the jelmeter, pectin will need to be added to the juice if a satisfactory gel strength is to be reached.

The other test for the adequacy of the pectin content of the juice itself is to stir together a tablespoon of the extracted juice and a tablespoon of denatured alcohol. This mixture will form a jelly-like mass if the pectin content of the juice is high enough for making jellies without added pectin. It is imperative that the test sample containing the denatured alcohol be thrown away without tasting. The equipment in contact with the alcohol also must be washed well. These precautions are necessary because the alcohol is dangerous when consumed. If the alcohol test does not produce the jelly-like mass, pectin will need to be added to the recipe.

In recipes where pectin is to be added, it is important to note the difference in procedure to be followed when using powdered pectin versus the sequence to be used when liquid pectin is being added. Powdered pectin is mixed initially with the cool fruit juice, and then the mixture is heated. However, liquid pectin is withheld until the juice and sugar mixture has been heated to a full boil.

When jellies are made without added pectin, the pectin concentration is somewhat lower than that provided in the jellies in which pectin is added. In order to reach the desired gel structure, it is necessary to concentrate the pectin in the juice by boiling off some of the extra liquid to achieve a somewhat smaller volume of jelly from the batch, but a jelly which has a gel structure. To accommodate this loss of total volume, the sugar level in the original formulation needs to be lower than would be used if pectin were being added. By using a lower initial level of sugar, it is possible to reach the appropriate concentration of sugar at the same time that the desired pectin concentration is obtained. Balances such as this between sugar, pectin, and acid must be achieved if the jelly is to have the desired characteristics.

Another modification recommended when a jelly without added pectin is being made is to use underripe fruit for about a fourth of the total

*Figure 9.15. Juices canned commercially are very convenient to use in making jellies of excellent quality.*

*Figure 9.16. Powdered pectin needs to be mixed with cold liquid prior to heating in order to achieve good distribution of the pectin and effective rehydration.*

*Figure 9.17. Liquid pectin can undergo hydrolytic breakdown and lose some of its gel-forming ability if it is heated too long. Consequently, liquid pectin is added after the fruit-sugar mixture has reached a boiling temperature.*

measure to help increase the pectin content from the fruit and to select ripe fruit for the remaining quantity of fruit.

**Boiling the jelly**

Jellies are boiled to help insure solubilization of the sugar and proper concentration of the pectin and sugar. Boiling needs to be done so that the entire mixture is so hot that the boil cannot be halted by stirring. The reason for such an aggressive state of boiling is that the necessary concentration of pectin and sugar needs to be reached without any unnecessary breakdown of the pectin and caramelization of the sugar. When a pectin-containing mixture is being heated, there is a slow breakdown of pectin to pectic acid. Although a small amount of this chemical change can be tolerated, it is not appropriate to permit extensive breakdown by very slow evaporation of the extra liquid in the jelly. Such a development would lead to jellies which either are too soft or which are very poor in color.

The recommendation of boiling jellies with added pectin for a full minute after the very hard, rolling boil is reached is a time block which needs to be monitored very carefully. With the very rapid rate of evaporation engendered by the vigorous boil, even a few seconds extra will cause the jelly to be too concentrated both in pectin and in sugar. This is the reason a clock equipped with a second hand should be used for timing the active boiling period of one minute.

**Testing for doneness**

Timing is the device used to determine doneness when preparing jellies with pectin added. However, timing is not a satisfactory answer for preparing jellies where no pectin has been added. Since the pectin content of the juices being used for making jellies without added pectin will vary considerably, there is no uniform time guideline available for establishing when the jelly will reach the necessary concentration of pectin. Instead, various tests have been devised. The use of a candy thermometer to determine when the temperature of the boiling jelly has reached a level 8°F higher than the temperature of boiling water at that location is one test. This measurement actually is measuring the concentration of sugar, but this level parallels the level of pectin concentration achieved. With this technique, it is essential that the appropriate temperature adjustment be made for locales at higher elevations.

*Figure 9.18. A rolling boil speeds evaporation of water from the product, thus minimizing the time available for breakdown of the sugar and pectin in the mixture.*

A time-honored test for doneness of jellies made without added pectin is the sheeting test. This test requires a visual assessment of the viscosity of the boiling jelly. This is a subjective type of test which requires some practice to be at all helpful. The test is conducted simply by dipping a cool, metal spoon into the boiling jelly and then tipping the spoon (out of the steam rising from the kettle) to note the flow characteristics of the liquid. An undercooked jelly will be light and rather free-flowing when the spoon is tilted. As the syrup is boiled and becomes more concentrated, the viscosity of the jelly increases until the drops coming from the tilted testing spoon flow together into a sheet as they fall from the spoon.

Yet another means of detecting the degree of doneness in boiling jelly mixtures is to put a little of the boiling jelly mixture on a plate and immediately place it in the freezer for a few minutes to determine whether a gel forms. While this test is being conducted, the jelly is removed from the heat so that it will not continue to cook until the need for additional cooking has been established. If the jelly is not done, the kettle is returned to the heat to continue the preparation of the jelly. Subsequent freezer tests are conducted until the jelly is determined by the test to be done.

Both the sheeting test and the freezer test require subjective assessments of the degree of doneness. Because of this, there is room for detectable errors to be made and for rather wide variations in quality of the finished product. Although these tests do provide some guidance, a more accurate result is obtained with the thermometer test. This statement assumes that the individual performing the test has compensated for any altitude adjustment needed.

**Filling the glasses**

As soon as the jelly has been boiled to the correct endpoint, the very hot jelly is ladled carefully, but quickly into the hot glasses. This step is important to minimize the tendency to disrupt the gel which will start to form fairly quickly. If the gel is disrupted after it is forming, the jelly will not be as clear as it will be if it is poured immediately after being removed from the heat. This quick packing into hot jars also is a useful means of reducing the likelihood of microbiological contamination in the jars or glasses. This latter aspect is of significance because, unlike the procedure used in canning, this open kettle method of preparing jams and jellies does not include heat treatment subsequent to packaging. The hot jelly is added to fill jars or glasses to within an eighth of an inch of the top of the container.

### Sealing the jelly

The packaged jellies need to be protected from contamination from the air. Such protection can be provided by sealing the jar if a regular canning jar is used or by sealing with paraffin. Either method works very satisfactorily when done correctly. The important matter is that there be no avenue left for air to penetrate to the jelly.

If a regular closure is to be used, the rim of the jar is wiped clean, and then the lid and band are placed in position and tightened firmly by hand,

*Figure 9.19. Just prior to ladling the jam or jelly into the glasses, the foam is skimmed from the surface. Elimination of the foam improves appearance and reduces the likelihood of microorganismal growth because of air trapped under the paraffin.*

*Figure 9.20. By transferring the very hot fruit mixture into very hot glasses, microorganismal contamination is reduced significantly.*

*Figure 9.21. The edges of the jars need to be wiped clean right down to the surface of the jelly to enable the paraffin to seal well.*

*Figure 9.22. If jars are to be heat processed (as is true for this apple butter) or are not being sealed with paraffin, a traditional canning closure is required for protection.*

*Figure 9.23. Hand tightening of the rings is done for closing jams and jellies, just as is done for canning other products.*

as is done in canning. If porcelain-lined zinc lids and rubber rings are being used, the rubber ring is moistened thoroughly before being placed on the shoulder of the empty jar. This placement actually is done before the jars are filled. Any spills will need to be wiped off before the lid is screwed on and tightened by hand. If the jars are packed and closed quickly, there should not be a problem in achieving a good seal. However, it is necessary to check the seal in the jelly jars in the same manner as that used for checking the seals of canned goods.

For jars to be sealed with paraffin, the level of jelly should allow half an inch of headspace at the top of the glass. Hot paraffin is poured immediately onto the surface of the jelly to just a shallow depth (1/8 inch thick). This thin layer of paraffin enables the paraffin to be more adaptable to stress as the jelly cools, thus helping to avoid breaks in the paraffin covering. The depth of the paraffin is a bit difficult to see when the melted paraffin is being poured onto the jelly. It is helpful to make a mental note of the point on the glass where the upper surface of the paraffin will come when the correct thickness is reached. To help avoid holes in the paraffin as it hardens, make very small air holes in the hot paraffin. These will close gradually as the paraffin flows back together on the hot jelly.

Melted paraffin represents a significant potential hazard because it can ignite and cause serious burns. The keys to avoiding problems with hot paraffin include the following practices:

1. Melting the paraffin in a double boiler or in a can placed in a pan containing simmering water. It definitely is unwise to place the paraffin in a pan over direct heat because of the risk of overheating the paraffin and causing a fire to break out.
2. Maintaining the paraffin at a temperature well below the smoke point of the material.
3. Always handling the container of paraffin carefully with hot pads. This greatly reduces the possibility of burning one's hands and spilling hot paraffin.
4. Keeping a large piece of heavy aluminum foil within easy reach of the range so that it can be dropped over the pan of paraffin if a fire should happen to break out. The aluminum foil effectively smothers the fire. With careful adherence to the recommendations given in the preceding points, there should never be a need to actually implement this suggestion. However, it is an important practice to maintain a state of readiness.

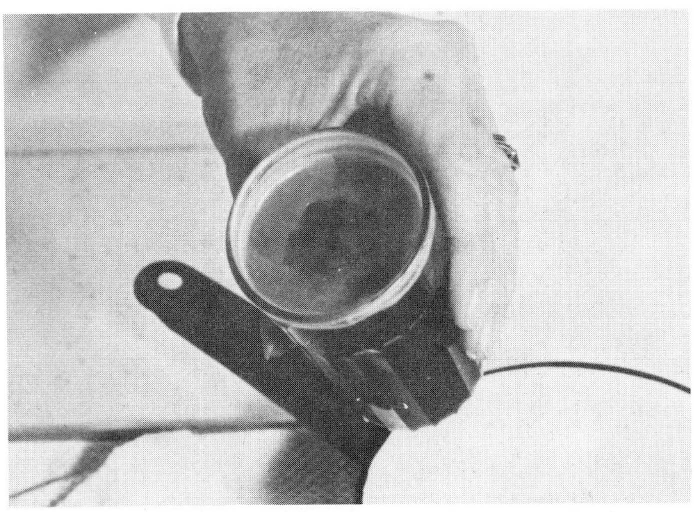

*Figure 9.24. A thin layer of paraffin is applied immediately to the surface of the hot jam or jelly as soon as the inside surface of the jar has been wiped clean. The jar should be tipped slightly and rotated to insure that the paraffin has flowed to effect a good seal all the way around the jar. Air must be prevented from entering through breaks in the paraffin.*

**Storing the jellies**

As the jelly begins to cool and form a gel initially, the structure of newly-prepared jellies is delicate. However, as the jelly rests without disturbance for half a day or more, the structure will begin to strengthen as more hydrogen bonds form. To obtain maximum strength of the gel and good clarity, jellies are allowed to remain on the counter without disturbance overnight. Then the glasses are covered with a metal lid, aluminum foil, or other suitable protective covering in preparation for storage.

Prior to being placed in long-term storage, the jars of jelly are labeled with the name of the product and the date of preparation. This enables a product to be identified and used prior to the end of a year. Storage should be in a cool and dark place; the temperature should undergo as little fluctuation as possible so that contraction and expansion of the paraffin will be minimized during storage. This also aids in maintaining good gel structure.

## Jams

Jellies are more time-consuming to make than jams because of the need to extract the juice. Since jams contain small pieces of the fruit, the fruit needs to be cut into pieces of the desired size, but the extraction of the juice is not done. The prepared fruit can be combined with pectin to make jams; powdered pectin is added to the cold, crushed fruit, just as the cool juice was combined with powdered pectin in making jellies. The liquid pectin is added to the jam immediately after the mixture has been boiled one minute. This means that the liquid pectin is not boiled at all, but instead is added and stirred in just prior to ladling the product into glasses or jars.

If jams are made without added pectin, the sugar and pectin are concentrated by the prolonged cooking period comparable to that required in making jellies without pectin being added. Again the preferred test is a thermometer reading. However, the freezer test also can be used. Jams are quite viscous and need to be stirred throughout their cooking period to avoid having pieces stick and burn.

## Preserves

Preserves are prepared in a manner very similar to that used in making jams. Fruits are prepared for preserves by washing them well, removing any stems and seeds, and then either leaving them whole or cutting into large pieces.

If preserves are being made without added pectin, the ingredients are boiled together until the temperature reaches 9°F. above the temperature of boiling water. It is necessary to stir the mixture during the boiling period to prevent sticking and burning and to help achieve uniform heating in the entire mixture.

For preserves made with added pectin, the method of preparation is dependent on the type of pectin being used. Powdered pectin is stirred into the cold, uncooked fruit prior to heating. The mixture is brought to a boil and then the sugar is added. As soon as the boiling is resumed throughout the preserves, the timing begins for a one-minute boiling period.

If jams and preserves are to be made with liquid pectin as a part of the recipe, preparation is begun by combining the fruit and sugar before heating these ingredients quickly to a boil. The mixture of fruit and sugar is boiled hard for one minute, with stirring being continuous to avoid burning and sticking. At the end of one minute of boiling, the kettle is removed from the heat, and the liquid pectin is stirred in well. Any foam on the surface is removed, and the preserves are placed in containers in preparation for sealing.

## Marmalades

Marmalades are made utilizing the peel of citrus fruits. Preparation of the peel includes cutting it into thin slices and then simmering until tender. The membranes and seeds are removed from the citrus fruits themselves before the pulp is added to the peel. The fruit and sugar are cooked together rapidly until the temperature reaches 9°F above the boiling temperature of water. It is necessary to stir continuously during the cooking period.

Since marmalades always are made using citrus fruits and citrus fruits are high in pectin naturally, pectin need not be added to a marmalade recipe. If pectin were added, the marmalade could be too viscous and might not have the desired spreading characteristics.

## Conserves

Conserves are made in the same manner as preserves and marmalades. However, the ingredients list is expanded by the addition of nuts and sometimes raisins. As is true with preserves, conserves may be made with or without added pectin. The procedures are the same as outlined above. However, the nuts are added after cooking has been completed so that their texture will be optimal.

## Uncooked Jams and Jellies

A variation of the conventional jams and jellies is provided by the preparation of uncooked products. These gels are distinctly less rigid than the gels resulting from the cooked products because the pectin and sugar concentrations are lower than are obtained by boiling the mixture. However, the color of these products is lighter and more comparable to the fresh or frozen fruit from which they are made than are the cooked items. The flavor also more closely resembles the fresh fruit than does the flavor from cooked jams and jellies.

## SUMMARY

Pectin gels are colloidal dispersions in which the carbohydrate, pectin, is the continuous solid phase and water is the discontinuous liquid phase. The essential components of pectin gels are pectin, sugar, and fruit juice (acid). The pectin provides the essential structural framework. Sugar contributes flavor and also assists in forming the gel structure by helping to dehydrate the pectin and by binding some of the water. The acid aids in reducing the charge on the pectin, thus helping to promote hydrogen bonding between pectin molecules to form the structural framework of the gel.

Jellies are pectin gels made from the extracted juice of barely ripe fruits. Jams are made from fruits cut into rather small pieces. Preserves contain either whole fruit or large pieces of fruit; conserves are jams or preserves with nuts and sometimes raisins added. Marmalade is the term applied specifically to citrus gels.

Pectin gels are prepared using the open kettle method of canning. This method simply involves boiling the mixture to a sugar concentration of 65 percent (9°F above the boiling temperature of water) and then packing the product in hot, clean jars which then are sealed with paraffin or with canning lids. The high concentration of sugar and the acidic nature (pH of approximately 3.3) of these products make them inhospitable to microorganisms. Therefore, heat processing following packaging is not necessary for safe storage. For optimum quality, it is recommended that pectin gels be consumed within a year. Some pectin gels are made with lower concentrations of sugar. These products do not have a high enough concentration of sugar to enable them to be stored at room temperature for long periods, but they are stored satisfactorily in the freezer.

Pectin gels may be made with added pectin; this pectin can be in either liquid or powdered form. The powdered form is added before boiling so that it will be rehydrated, whereas the liquid form is added after cooking is completed to avoid breaking the pectin into the non-gelling pectic acid. Jams and jellies may be made without added pectin simply by boiling long enough to evaporate some of the liquid and concentrate the pectin enough to gel the mixture.

## BIBLIOGRAPHY

Baker, G.L. High-polymer pectins and their deesterification. *Adv. Food Res. 1:* 395. 1948.

Ball Corp. *Blue Book.* Ball Corp., Muncie, Ind. 29th ed. 1972.

Consumer and Food Economics Institute. *How to Make Jellies, Jams, and Preserves at Home.* Home and Garden Bulletin No. 56. U.S. Dept. of Agriculture. Washington, D.C. Rev. 1975.

Joslyn, M.A. Chemistry of protopectin: a critical review of historical data and recent developments. *Adv. Food Res. 11:* 2. 1962.

Kerr Glass. *Kerr Home Canning and Freezing.* Kerr Glass Manufacturing Corp. Sand Springs, Okla. 10th ed. 1975.

Kertesz, Z. I. *Pectic Substances.* Interscience Publishers. New York. 1951.

Neubert, A.M. and J.L. St. John. Fruit juice, jams, jellies, and preserves. Chapter 44 in *Chemistry and Technology of Food and Food Products.* Ed. by M.B. Jacobs. Vol. 3. 2nd ed. Interscience Publishers. New York. 1951.

# Chapter 10.
# Jams, Jellies, & Preserves

## EQUIPMENT

### Basic equipment

An essential piece of equipment in making jams, jellies, and preserves is a large kettle so that it is possible to boil the mixture hard without worry about needing to reduce the heat to keep it from boiling over. A jelly bag and colander are helpful adjuncts for expressing the juice when making jellies, but these are not needed for other pectin-containing products. For products made without added pectin, a candy thermometer is needed so that the concentration of sugar can be noted with accuracy. Other useful items include long-handled wooden spoons, a ladle, a clock with a second hand, and a potato masher. When making jellies without added pectin, a jelmeter is a useful guide to the appropriate amount of sugar to add in relation to the pectin contained in the extracted juice.

### Containers

Jelly glasses or jars need to be washed thoroughly in hot, soapy water and then rinsed completely in hot water. The containers then should be boiled for 10 minutes to sterilize them. They can then be kept hot in an oven set at 200°F. or in hot water until they are used. This measure and the sterilizing operation are important to the packaging of these products. The finished, hot jam or jelly should be placed in a sterile environment to reduce the risk of spoilage. This sterilizing is important here because, unlike canning, jams and jellies usually are not submitted to heat processing after they are transferred into the container in which they will be stored.

The lids and bands also need to be washed and rinsed well. They then should be treated in the manner suggested by the manufacturer. New lids should be used unless closures are being used which are designed for repeated use or paraffin is being used as the sealing material.

### Paraffin

Paraffin represents a potential risk in the making of jams and jellies because it ignites readily. It is essential to heat the paraffin in a can or pan placed in a shallow pan of water (simulating a double boiler). Paraffin should be heated with a slow rate of heating. The can containing the paraffin should be handled with hot pads to reduce the likelihood of dropping the hot paraffin.

For best results when sealing with paraffin, a layer only about ⅛ inch deep should be poured on the surface of the product. This is thick enough to provide a complete seal over the surface of the food, yet is thin enough to allow for expansion and contraction with temperature changes. It also is important to eliminate any air pockets in the layer by inserting a fork tine or toothpick into any bubbles in the paraffin

## PROCEDURES

### Use of Pectin

The firmness of jams and jellies is influenced by the amount of pectin, the concentration of sugar, and the acidity of the mixture. The pectin content of extracted juices can be determined by measuring the viscosity of the juice in a jelmeter. If the pectin content is low, the juice will drain quite quickly from the jelmeter and will indicate that a very limited amount of sugar should be added so that the sugar will not become too concentrated in the evaporating mixture before the pectin content reaches the necessary level. Conversely, very high pectin content causes a juice to drain slowly, indicating that a proportionately larger amount of sugar should be added because only a short boiling period will be needed to achieve the desired concentrations. Since the concentration of sugar can be traced by elevation of the boiling point, a thermometer can be used to indicate the correct endpoint.

Although some fruits have enough pectin in them to gel the jams and jellies, many of them do not contain enough pectin. To remove uncertainties about the outcome of pectin gels, it is possible to add pectin from a commercial source. The powdered pectins require rehydration during preparation of the product and, consequently, the usual practice is to add the powdered pectin prior to the cooking period. Liquid pectin, however, is marketed in the hydrated form and, as such, is more susceptible to chemical hydrolysis during the boiling period than is the powdered form. For this reason, liquid pectin is stirred in thoroughly following the boiling period. Some of the thickening ability of the pectin would be lost if liquid pectin were allowed to undergo hydrolysis.

## Open Kettle Canning

Jams, jellies, and preserves ordinarily are made by a method known as *open kettle canning.* In this method, the food is cooked to the desired degree in an open kettle before it is placed in sterile jars. This is in contrast to the procedures used in canning fruits, vegetables, meats, poultry, and fish. In these items, the required heat processing in the jar is appropriate in order to make the food safe for storage, whereas open kettle canning simply is unsafe for such foods. The open kettle canning method can only be used in jams, jellies, and preserves because of the safety provided by the very high sugar content and the acid.

## Water Bath Canning

When butters are being prepared, these products must be processed in a water bath canner in boiling water to insure good shelf storage. This is necessary because the concentration of sugar is very low. A pressure cooker is not required; the acidity of the medium is sufficient to make water bath canning effective.

When making jams and jellies which will be stored under warm or humid conditions, processing in a water bath canner is recommended even when the jams and jellies contain the usual, high amounts of sugar. The recommended processing is gentle compared with the treatments used on some canned produce. The suggested time for processing with the water actually boiling is five minutes.

## Storage

Traditional jams, jellies, preserves, and conserves can be stored at room temperature if they are protected by a thin layer of paraffin or if they are in sealed canning jars. The somewhat less stringent precautions used in processing and storing these products, in comparison with canned goods, is the result of the protective quality of the sugar due to the unfavorable medium it presents for growth of microorganisms.

Some special pectin products are made using a lower concentration of sugar. Since these have less protection from spoilage because of the reduced sugar concentration, refrigerator or freezer storage is necessary to avoid harmful growth of microorganisms. The disadvantage of the need for this special storage is apparent, but this is considered by some people to be outweighed by the advantages of flavors and colors more nearly approximating the fresh fruit. These advantages accrue because it is not necessary to boil the product to concentrate the pectin and sugar.

# EVALUATION

Jellies should be firm enough to hold their shape, yet malleable enough to spread easily. The product is translucent to transparent, and it has a pleasing color and full flavor. The texture is smooth, but there is a slight resistance to pressure in the mouth. There will be no suggestion of graininess. Jams have several characteristics which are similar. The most apparent difference is the more opaque quality which results from the presence of the pieces of fruit. This same situation is true for preserves and conserves. Despite the presence of the fruits, there still is a translucent quality to the well-prepared finished product.

Sometimes pectin gels are observed to have a rather cloudy appearance. This results when the pectin gel is disturbed after it begins to set in its final form. This problem is avoided by ladling the product into the jars before there has been sufficient cooling of the hot mixture to allow the definition of the structure.

Crystals will form in jams and related products if the concentration of sugar is too high. This situation ordinarily is the result of boiling the product to too high a final temperature. Another possible cause of crystals can be seen in grape products. The tartaric acid in grapes (which serves also as the basis of cream of tartar) crystallizes readily in the form of tartrate salts, thus contributing to the gritty texture sometimes noted.

Some jellies and jams fail to hold their desired shape after the structure has had time to set. This can be due to failing to achieve the desired concentration of pectin and sugar. If there is too little acid, pectin, or sugar, the item may be too soft. The slow rate of heating resulting from making very large batches also can cause the jam to be too soft.

This is explained by the fact that there is considerable hydrolysis of the carbohydrate during boiling and this reduces the gel strength.

Jams and jellies which are too soft can be remedied by bringing ¼ cup water and 4 teaspoons powdered pectin to a boil and then stirring in ¼ cup sugar and 4 cups of the soft jelly or jam. The product is returned to a boil and boiled vigorously for 1 minute before skimming and proceeding with the remaining packaging steps.

Soft jams and jellies also can be remedied using liquid pectin. In this product, the jelly is heated very rapidly to boiling. Then ¾ cup sugar, 2 tablespoons lemon juice, and 2 tablespoons liquid pectin are stirred in. Boiling is continued for 1 minute before the jelly containing the liquid pectin is removed from the heat, skimmed, and packaged according to the usual procedures for covering, labeling, cooling, and storing. It also is possible to reboil a jam or jelly without adding pectin. The easiest way to be certain the correct endpoint is achieved with this treatment is with the use of a thermometer (boil until temperature is 9°F above the boiling point of water).

Some jams or jellies are too stiff rather than too thin. This condition results when the pectin concentration is too high. This high concentration may be the result of using too much pectin in the recipe or of cooking the product to too high a concentration of pectin. If carried to an extreme, overcooking will cause jams and jellies to become rather gummy and rubbery.

## JELLIES

### Green Pepper Jelly

2 long green chili peppers (about 4 inches long)
1 medium green bell pepper
1½ cups cider vinegar
5 cups sugar
Few drops of green food coloring
½ cup liquid pectin

1. Wash, seed, and cut peppers in large pieces.
2. Whirl ½ cup of the vinegar and the peppers in the blender to puree.
3. Pour into a large kettle. Rinse the blender with the remaining vinegar and add to the kettle.
4. Add the sugar to the kettle, and slowly heat mixture to boiling while stirring constantly to help dissolve the sugar.
5. Boil hard for 2 minutes.
6. Remove from heat. Skim off the foam before straining the pepper mixture through cheesecloth.
7. Add green food coloring to obtain a green approximately comparable to the fresh chili color.
8. Stir in the liquid pectin, being sure to mix well.
9. Ladle in hot jars or glasses and seal with paraffin or canning closures before labeling and storing.

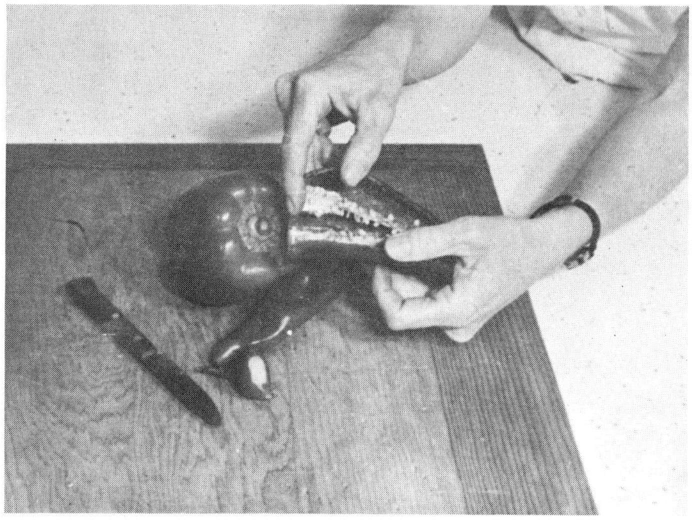

*Figure 10.1. Preparation of Green Pepper Jelly begins with the removal of the seeds from the washed peppers. The seeds of the chili peppers would create a burning sensation in the mouth if they were not removed.*

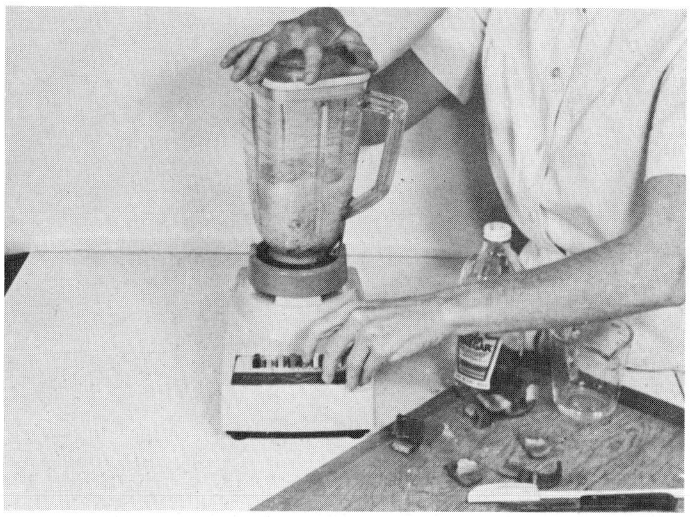

*Figure 10.2. Pieces of the peppers are whirled in the blender with part of the vinegar until pureed.*

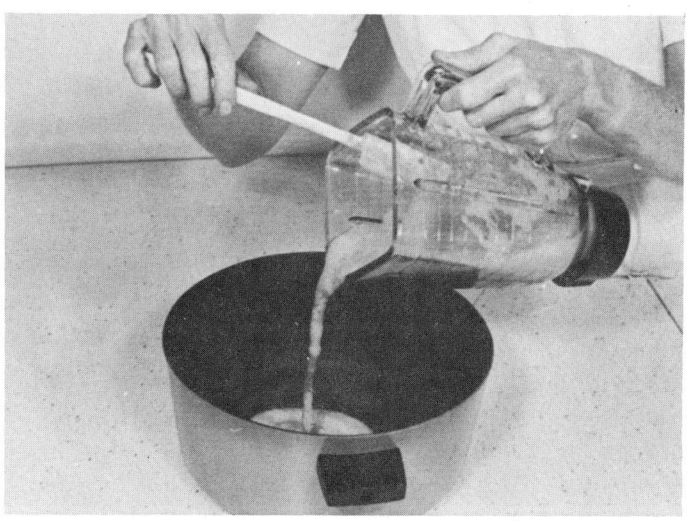

*Figure 10.3. The pureed peppers are poured into a large kettle preparatory to boiling of the jelly.*

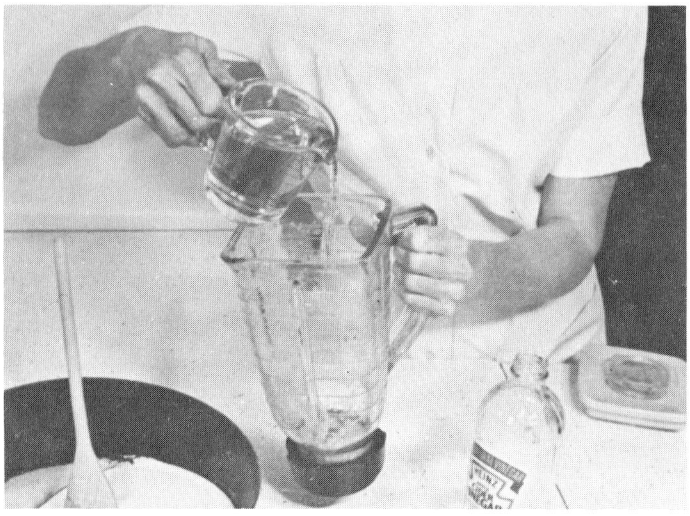

*Figure 10.4. The remainder of the vinegar is used to rinse all remaining pepper puree from the blender jar.*

*Figure 10.5. The sugar is added to the pepper-vinegar puree.*

*Figure 10.6. To avoid burning the jelly, the mixture is stirred continuously while boiling is continued for two minutes at a hard boil.*

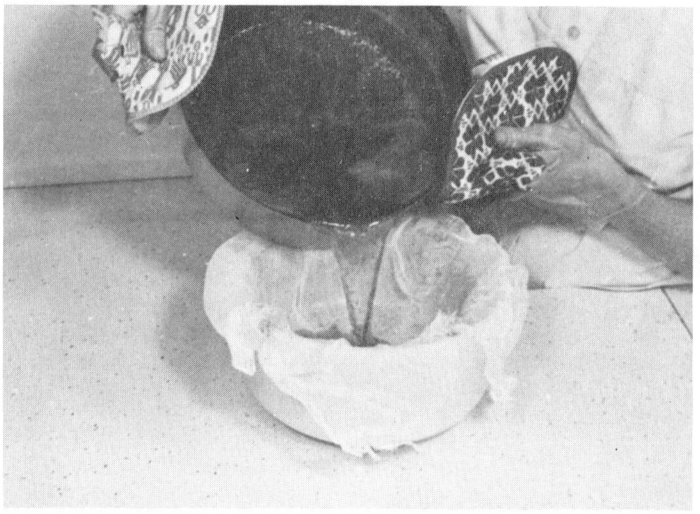

*Figure 10.7. Cheesecloth arranged in a double layer in a metallic, perforated basket is used to strain the juice from the pepper puree after the mixture is removed from the heat.*

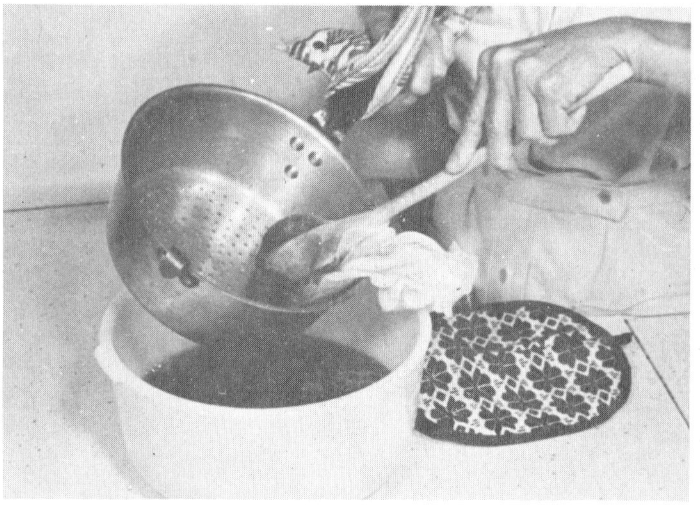

*Figure 10.8. As much liquid as possible is pressed from the cheesecloth.*

*Figure 10.9. Green food coloring is added, drop by drop, to attain the desired green color of the fresh green peppers.*

*Figure 10.10. Since liquid pectin is being used, the pectin is stirred in very thoroughly after the liquid has been strained through the cheesecloth. This assures maximum gelling from the pectin.*

*Figure 10.11. The completed jelly quickly is ladled into hot glasses in preparation for protection with paraffin and storage.*

*Note:* Excellent as a flavor accent with meat or as a jelly served with cream cheese on crackers for hors d'oeuvres.

## Cranberry Wine Jelly

3 cups sugar
1 cup cranberry juice
1 cup red wine
½ bottle liquid pectin

1. Mix sugar, cranberry juice, and wine together well in a large kettle.
2. Heat to a boil while stirring and continue to boil for 1 minute.
3. Remove from heat, skim off the foam, and stir the liquid pectin in carefully and thoroughly.
4. Ladle into hot jars or glasses and seal with paraffin or canning closures before labeling and storing.

*Note:* Serve this jelly as a meat accompaniment or with other entrees.

## Creme de Menthe Jelly

2½ cups sugar
1 cup water
½ bottle liquid pectin
1 six-oz. can lemonade concentrate (undiluted)
½ cup Creme de Menthe

1. Combine sugar and water in a saucepan and heat to a rolling boil, stirring constantly.
2. Boil hard for 1 minute before removing from the heat.
3. Stir in the pectin.
4. Mix the lemonade concentrate and Creme de Menthe into the sugar solution.
5. Ladle into hot jars or glasses and seal with paraffin or canning closures before labeling and storing.

## Apricot Muscatel Jelly

3½ cups sugar
1 cup apricot nectar
2 tablespoons lemon juice
1 cup Muscatel wine
½ bottle liquid pectin

1. Combine all ingredients except the pectin and heat to a boil while stirring constantly.
2. Boil the mixture hard for 1 minute.
3. Remove from the heat, skim off the foam, and stir the liquid pectin in carefully and thoroughly.
4. Ladle into hot jars or glasses and seal with paraffin or canning closures before labeling and storing.

## Apple Jelly (Canned Juice)

4 cups canned apple juice
5 cups sugar
1 box powdered pectin

1. Combine the apple juice and the powdered pectin in a large saucepan. Mix well.
2. Heat the juice and pectin to a boil while stirring constantly.
3. Add the sugar and continue stirring while heating the mixture to a rolling boil.
4. Boil the mixture hard for 1 minute.
5. Remove from the heat and skim off the foam before ladling into hot jars or glasses.
6. Seal with paraffin or canning closures before labeling and storing.

## Apple Jelly (Fresh Fruit)

8 lb. tart, ripe apples
13 cups water
15 cups sugar
1 bottle liquid pectin

*Figure 10.12. Powdered pectin requires rehydration during cooking of the jelly and, consequently, is combined with the juice at the beginning of the cooking period.*

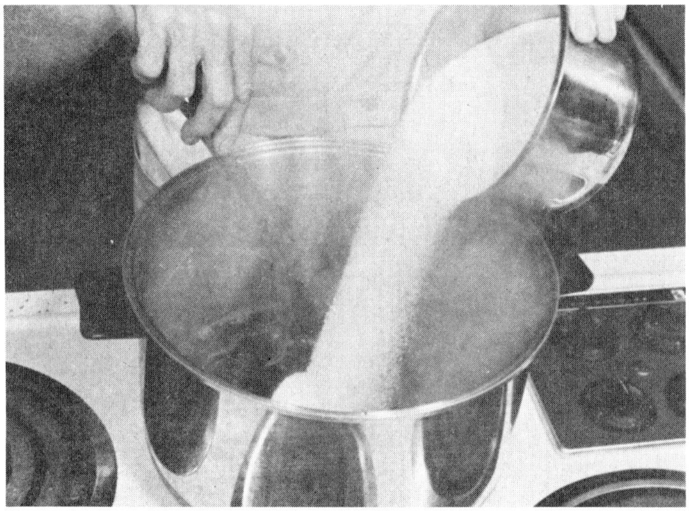

*Figure 10.13. After the juice and powdered pectin have been brought to boiling, the sugar is added. This later addition of the sugar reduces the length of time the sugar is heated in the presence of the acidic fruit, thus reducing the amount of hydrolysis the sugar may undergo.*

*Figure 10.14. After the powdered pectin-juice-sugar mixture returns again to a boil, the jelly is boiled hard for one minute to achieve the desired concentration of sugar and pectin. The jelly then is ready for skimming, ladling into jars, sealing, and storage.*

1. Quarter unpeeled apples, removing the stem and flower ends of each piece. Chop into small pieces without removing the core. The core is needed to add to the pectin in the jelly.)
2. Bring the apple pieces and water to a boil rapidly.
3. Simmer for 10 minutes.
4. Mash the apples with a potato masher and then continue simmering for an additional 5 minutes with the cover on the pan.
5. Place a couple of layers of cheesecloth or a jelly bag in a colander before transferring the fruit from the pan to the colander (placed over a wide pan or kettle to collect the juice).
6. Measure 10 cups of the juice into a large kettle and add the sugar.
7. Heat quickly to a boil.
8. Boil for 1 minute and then add the pectin while stirring.
9. Remove from the heat, and skim off the foam. Ladle into jelly glasses, seal, label, and store.

## Honey Jelly

2½ cups honey
¾ cup lemon juice
½ cup liquid pectin

1. Combine honey and lemon juice in a saucepan.
2. Heat to a rolling boil while stirring continuously. Boil hard for 2 minutes.
3. Remove from heat before stirring in pectin.
4. Ladle into hot jars or glasses.
5. Seal with paraffin or canning closures before labeling and storing.

## Grape Jelly (Fresh Fruit)

4 cups grape juice (about 3 lbs. ripe grapes)
7 cups sugar
½ bottle liquid pectin

1. Wash and stem the grapes and crush well. (Concord grapes are the best variety because of their color and flavor.)
2. Place the crushed grapes in a kettle and add ½ cup of water. Heat to boiling, covered, and then simmer for 10 minutes.
3. Place the grapes in a jelly bag and squeeze out the juice.
4. Measure the 4 cups of juice into a large kettle, and add the sugar.
5. Heat to boiling while stirring constantly.
6. Add the pectin, and boil hard for 1 minute.
7. Remove from the heat. Skim off the foam before ladling into hot jars or glasses.
8. Seal with paraffin or canning closures before labeling and storing.

## Grape Jelly (Bottled Juice)

4 cups bottled grape juice
7 cups sugar
1 bottle liquid pectin

1. Heat the juice and sugar to a boil while stirring.
2. Boil hard for 1 minute; stir in the liquid pectin.
3. Remove from heat, skim off foam, and ladle into hot jars or glasses.
4. Seal with paraffin or canning closures.
5. Label and store.

## Grape Jelly (Frozen Concentrate)

6½ cups sugar
2½ cups water
1 bottle liquid pectin
3 cans (6 oz. each) frozen grape juice concentrate

1. Heat sugar and water to boiling in a kettle, stirring constantly.
2. Boil hard for 1 minute.
3. Remove from heat before stirring in the liquid pectin.
4. Stir in undiluted, thawed grape juice concentrate.
5. Ladle into hot jars or glasses.
6. Seal with paraffin or canning closures before labeling and storing.

# JAMS

## Strawberry Jam (Fresh Fruit)

4 cups crushed strawberries (about 2 quart boxes)
7 cups sugar
½ bottle liquid pectin

1. Sort and wash fully ripe berries, being careful to remove stems and caps. Crush with a potato masher.
2. Measure the berries and sugar into a large kettle, stirring well.
3. Heat quickly to a hard boil while stirring constantly.
4. Boil hard for 1 minute.
5. Remove from heat and stir in pectin.
6. Skim off the foam.
7. Ladle into hot jars or glasses and seal with paraffin or canning closures.
8. Label and store.

Yield: 8 half pints

## Strawberry Jam (Fresh, Uncooked)

4 cups fresh strawberries
4 cups sugar
2 tablespoons lemon juice
½ bottle liquid pectin

1. Wash the berries well. Remove the caps and any spoiled or bruised areas.

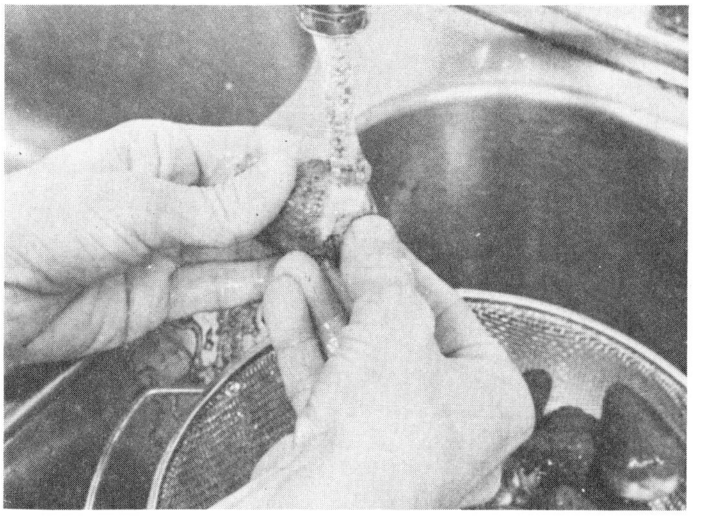

*Figure 10.15. Fresh strawberries are sorted carefully, washed, and capped preparatory to making jam.*

*Figure 10.16. After the fresh berries have been mashed with a potato masher, the sugar is stirred in.*

*Figure 10.17. The berry-sugar mixture is brought to a rolling boil in a large kettle while stirring constantly to prevent sticking and burning. After boiling hard for one minute, the jam is removed from the heat, and the liquid pectin is stirred in very thoroughly.*

*Figure 10.18. The foam on the surface of the jam is removed carefully, but quickly with a large cooking spoon.*

*Figure 10.19. The foam removal needs to be done as quickly as possible, yet completely so there will not be air bubbles on top of the jam which will interfere with the protective coating provided by the paraffin seal. Speed is required so that the hot jam will be more sterile when placed in the jar, and so the structure of the gel will not have started to form in the jam.*

*Figure 10.20. The jam then is ladled into hot glasses or jars.*

*Figure 10.21. Paraffin is applied immediately after the jam is placed in the jars. However, a decorative coating using whipped paraffin can be applied when convenient. Whipping of paraffin is facilitated by using a can large enough to hold the paraffin and the beater blades of a portable electric mixer. This can be stored until needed and then placed in a pan of hot water for melting the paraffin.*

*Figure 10.22. The melted paraffin is whipped just before it begins to set, but while it is still very malleable. Beating continues until a rather fluffy foam forms.*

2. Crush with potato masher.
3. In a bowl, combine the crushed fruit, sugar, and lemon juice. Stir together thoroughly.
4. Let the sugar-berry mixture sit for 20 minutes before adding the liquid pectin.
5. Stir mixture together for 3 minutes to distribute pectin absolutely uniformly.
6. Ladle into jars and cover tightly.
7. Store in refrigerator no longer than 3 weeks; for longer storage, place immediately into freezer.

*Note:* Powdered pectin may be used in place of liquid pectin by combining a package of powdered pectin with ¾ cup water, heating the mixture to a boil, and boiling for 1 minute. This mixture then is added to the berry-sugar mixture in exactly the same way as the liquid pectin.

*Figure 10.23. A small piece of aluminum foil is placed over the thin, protective layer of paraffin, and the foil is covered with the whipped paraffin. Decorative arrangement of the paraffin foam is done easily with the aid of a small spatula.*

*Figure 10.24. Decorative toppings can be trimmed with a variety of small bows or arrangements to suit the gift of jam to the recipient and the occasion.*

*Figure 10.25. Fresh strawberries can be used also to make an uncooked jam, which is similar to the fresh berries in color and flavor. Since the sugar content is not concentrated by boiling off some of the liquid, this jam requires refrigerator or freezer storage to prevent spoilage.*

## Strawberry Jam (Frozen, Uncooked)

2 packages (10 oz. each) frozen strawberries, thawed
3⅓ cups sugar
½ bottle liquid pectin
2 teaspoons lemon juice

1. Allow berries to warm to room temperature.
2. Stir in the sugar until it is dissolved completely. If necessary to dissolve all of the sugar, warm the berries very slightly.
3. Stir the pectin in thoroughly.
4. Ladle into jars and cover tightly.
5. Store in refrigerator no longer than 3 weeks. For longer storage, place directly into the freezer.

## Strawberry Jam (Frozen, Cooked)

2 packages (10 oz. each) frozen strawberries
3 cups sugar
⅓ cup lemon juice
¼ cup liquid pectin

1. Place berries and their syrup in a large kettle and partly thaw until blocks can be broken apart with a wooden spoon.
2. Add sugar and lemon juice, mixing well with a wooden spoon.
3. Heat to a rolling boil, and continue to boil hard for exactly 1 minute.
4. Remove from the heat, and immediately stir in the liquid pectin.
5. Skim off any foam, and ladle into hot jars or glasses.
6. Seal with paraffin or canning closures, label, and store at room temperature.

## Raspberry Jam (Frozen, Cooked)

2 packages (10 oz. each) frozen raspberries
3 cups sugar
⅓ cup lemon juice
¼ cup liquid pectin

1. Place berries and their syrup in a large kettle and partly thaw until blocks can be broken apart with a wooden spoon.
2. Add sugar and lemon juice, mixing well with a wooden spoon.
3. Heat to a rolling boil, and continue to boil hard for exactly 1 minute.
4. Remove from the heat, and immediately stir in the liquid pectin.
5. Skim off any foam, and ladle into hot jars or glasses.
6. Seal with paraffin or canning closures, label, and store at room temperature.

## Cherry Jam (Liquid Pectin)

4½ cups ground (finely chopped) pitted sour cherries
7 cups sugar
1 bottle liquid pectin

*Figure 10.26. Frozen raspberries (or other frozen fruit) are available any time of the year, making it possible to make excellent jams regardless of the time of year.*

*Figure 19.27. As soon as the berries, sugar, and lemon juice have been boiled hard for a minute and the liquid pectin has been stirred in, the foam is skimmed completely from the jam.*

*Figure 10.28. The skimmed jam is ladled into jars or glasses in preparation for sealing with melted paraffin. The wide-mouthed funnel is a convenient tool to fill the glasses without spilling the rather fluid jam on the glass.*

1. Sort, wash, and pit fully ripe cherries. Grind or chop fine.
2. Measure the cherries into a kettle, add sugar, and heat to boiling while stirring constantly.
3. Boil hard for 1 minute while stirring.
4. Remove from heat, and stir in the liquid pectin.
5. Skim foam from jam, ladle into hot jars or glasses, seal, label, and store.

Yield: 8 half pints

## Cherry Jam (Powdered Pectin)

4 cups ground (finely chopped) pitted sour cherries
1 package powdered pectin
5 cups sugar

1. Sort, wash, and pit fully ripe cherries. Grind or chop fine.
2. Measure cherries into a kettle, add the pectin, and stir well.
3. Place over high heat, and stir continuously while heating to a rolling boil.
4. Add the sugar while stirring and continuing to heat. Bring back to a boil and boil hard for exactly 1 minute.
5. Skim off the foam before ladling into hot jars.
6. Seal with paraffin or canning closures. Label before storing.

Yield: 6 half pints

## Peach Jam

4¼ cups crushed peaches (about 3½ pounds)
¼ cup lemon juice
7 cups sugar
½ bottle liquid pectin

1. Sort and wash fully ripe peaches. For convenience in peeling, dip very briefly in boiling water just long enough to loosen the skin.
2. Peel and pit the peaches. Crush with a potato masher.
3. Immediately measure crushed peaches into a kettle and stir in lemon juice and sugar.
4. Heat rapidly to a full boil, and boil hard for 1 minute.
5. Remove from heat, and stir in the liquid pectin.
6. Skim off any foam before ladling into jars or glasses, sealing, labeling, and storing.

Yield: 8 half pints

## Freezer Apricot Jam

2 cups peeled, pitted, mashed apricots
4 cups sugar
3 tablespoons lemon juice
½ cup liquid pectin

1. Combine apricots, lemon juice, and sugar, stirring thoroughly to help dissolve the sugar.
2. Let stand for 20 minutes, stirring occasionally.

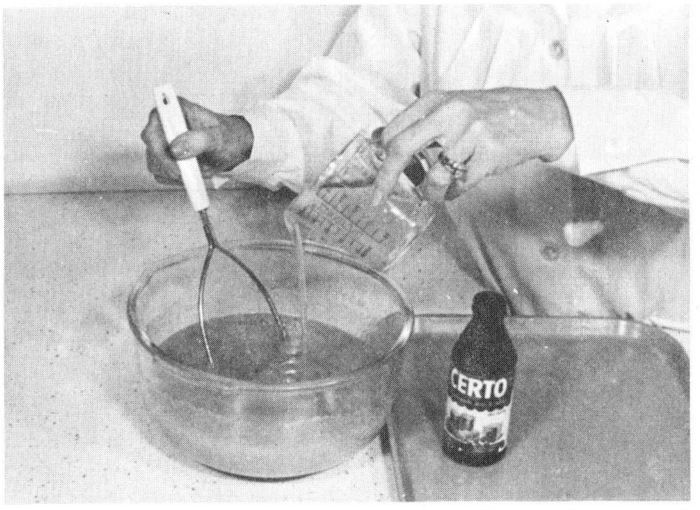


*Figure 10.29. Fresh apricots can be prepared as a freezer jam very quickly, a fact which will be appreciated if the crop is very abundant. The washed, pitted, and peeled fruit is mashed prior to the addition of the sugar and lemon juice. Mild heating may be needed to dissolve all of the sugar. This jam requires refrigerator or freezer storage.*

3. If mixture is still gritty from undissolved sugar, warm very gently while stirring. Avoid warming beyond the minimum amount needed to dissolve the sugar.
4. Remove from the heat (if mixture has been heated at all); stir in the pectin thoroughly for 2 minutes.
5. Immediately pour into clean jars and cover with clean lids or other air-tight cover. Let stand at room temperature for an hour.
6. Store in the refrigerator for no longer than 6 weeks or in the freezer for up to a year.

Yield: 5 half pints

## PRESERVES

### Strawberry Preserves

6 cups whole, capped strawberries
4½ cups sugar

1. Wash and cap firm, large, tart strawberries.
2. Combine fruit and sugar in alternate layers and refrigerate for 10 hours to draw out some of the juice from the berries and dissolve the sugar.
3. Heat quickly to boiling in large kettle while stirring.
4. Boil hard until temperature reaches 9°F. above the boiling point of water or until the preserves become somewhat thick (about 15 minutes).
5. Skim the foam as soon as the preserves are removed from the heat; ladle into hot, clean glasses or jars.
6. Cover with ⅛ inch of paraffin or with canning closures.
7. Label and store.

Yield: 4 half pints.

## CONSERVES

### Versatile Conserves

1 package (8-oz.) dried apricots, peaches, prunes, pears, or mixed fruit
Water
2½ cups (1 No. 2 can) crushed pineapple
⅔ cup seedless raisins
½ cup finely chopped nut meats
2 tablespoons lemon juice
3 cups sugar
½ bottle liquid pectin

1. Soak dried fruit in water to cover for at least 4 hours.
2. Drain. Chop rehydrated fruit.
3. Combine the rehydrated, chopped fruit with the undrained pineapple, raisins, nuts, lemon juice, and sugar in a large kettle, stirring to mix very well.
4. Stir constantly while bringing mixture to a boil and also during the 1 minute of hard boiling.
5. Remove from heat, and stir in the pectin thoroughly.
6. Skim the foam and stir slowly between 5 and 7 minutes to allow enough cooling so that the fruit will not float to the surface of the conserves.
7. Ladle into hot, clean glasses or jars and cover with ⅛ inch of paraffin or with canning closures.
8. Label and store when cool.

Yield: 5 to 6 half pints

### Apple Conserve

4½ cups finely chopped tart red apples (3 lb.)
½ cup water
¼ cup lemon juice
½ cup raisins
1 package powdered pectin
5½ cups sugar
½ cup chopped walnuts
1 tablespoon grated orange rind

1. Wash apples well; cut in quarters and remove stem and blossom ends and core. Do not pare. Chop apples into small pieces.
2. Combine apples, water, orange rind, lemon juice, raisins, and pectin in a large kettle; stir very well.
3. Heat rapidly to a full, rolling boil. Be sure to stir constantly.
4. While still on the heat, stir in the sugar and boil hard for 1 minute after mixture returns to a boil. Stir constantly.
5. Remove from the heat and immediately stir in the nuts.
6. Skim the foam before ladling conserves into jars or glasses and sealing (with paraffin or canning closures), labeling, and storing.

*Note:* A very small amount of red food coloring can be added if a more intense color is preferred.

# MARMALADE

## Orange Marmalade

6 navel oranges
4 lemons
¼ teaspoon baking soda
3 cups water
10 cups sugar
½ bottle liquid pectin

1. Cut the skin on the oranges and lemons into quarters and peel the fruit.
2. Remove about half of the albedo (white portion of the skin). Slice the remaining rind into very thin slices or shavings.
3. Add the water and soda to the rind. Simmer in a covered saucepan for 20 minutes, stirring occasionally.
4. Meanwhile, remove seeds from the pulp and cut pulp into small pieces.
5. At the end of the 20-minute simmering period, add the pulp to the rind and simmer 10 more minutes.
6. Transfer 6 cups of the fruit mixture to a large kettle, stir in the sugar, and heat rapidly to a rolling boil while stirring constantly and continue to boil hard an additional minute.
7. Remove from the heat and immediately stir the pectin in thoroughly.
8. Skim and stir the mixture alternately for 7 minutes before ladling into hot, clean jars or glasses.
9. Seal, label, and store.

Yield: 12 half pints

*Figure 10.30. Orange marmalade is made utilizing slices of the fruit, including the rind. The rind contributes flavor as well as pectin. Although it is possible to make marmalade without added pectin, the use of some commercial pectin helps to insure gelling without needing to use an amount of rind which could cause bitterness in the marmalade.*

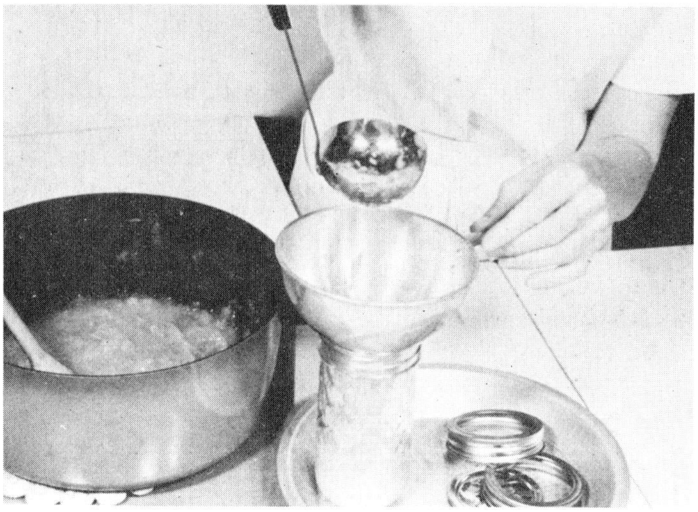

*Figure 10.31. The cooking of marmalade involves a period to help release pectin from the rind and also to concentrate the sugar in the mixture. After the liquid pectin has been added and stirred in, the marmalade is ladled into the jars and sealed.*

## Rhubarb Marmalade

3 pounds rhubarb
4 tablespoons shredded orange rind
1½ cups water
11 cups sugar
1 bottle liquid pectin

1. Thinly slice well-washed, unpeeled rhubarb.
2. Heat rhubarb, rind, and water to simmering in a saucepan; simmer until tender (about 1 minute).
3. Place 6 cups of the fruit and the sugar into a large kettle and bring quickly to a boil while stirring constantly.
4. Boil hard for 1 minute while stirring.
5. Remove from heat; stir in the pectin thoroughly.
6. Continue alternately stirring and skimming the foam for 5 minutes before ladling into hot, clean jars or glasses.
7. Seal with paraffin or canning closures, label, and store.

Yield: 10 half pints

# BUTTERS

## Apple Butter

8 pounds firm, red apples (McIntosh, Jonathan, or Winesap)
2 cups apple juice or cider
8 cups sugar
1 cup cider vinegar
1 teaspoon salt
1 teaspoon ground cinnamon
1 teaspoon ground cloves
¼ cup sugar
½ teaspoon ground ginger (optional)
½ teaspoon ground allspice (optional)
1 teaspoon grated lemon rind (optional)

1. Wash, quarter, and core, but do not peel apples.
2. Pour ¼ cup juice into the blender and gradually add apple chunks, blending between each addition, until about 2 cups of pureed apple are in the blender. Transfer puree to a large, shallow pan (such as the bottom of a roaster), and repeat the process of adding ¼ cup juice to the blender and gradually adding the apples to make a second 2-cup portion of pureed apple. Repeat the process until a total of 16 cups of pureed apples has been prepared.
3. In the roaster pan, stir together the apple pulp, 8 cups of sugar, and the vinegar. Cover very loosely with aluminum foil and bake for 2 hours at 325°F.
4. Mix the remaining ingredients together and stir into the hot mixture after it has completed the 2-hour baking period.
5. Again loosely cover the pan and continue baking another hour until the consistency is viscous enough to spread without running. This point also is tested by putting a tablespoon of the hot apple butter on a plate to determine whether liquid separates readily. If no rim of liquid is formed, the butter is ready for canning.
6. Ladle the hot butter into hot, clean canning jars, leaving ½ inch headspace.
7. Wipe the rims, close the jars with canning closures according to manufacturer's directions, and process in boiling water in a water bath canner for 10 minutes after the water is boiling actively in the canner.
8. After processing is completed, remove from the canner, cool, check seals, label, and store as for other canning.

*Note:* Water bath canning is required for the processing of apple and other butters because the sugar content is too low to provide the preserving action it contributes to other pectin-containing products.

## NON-SUGAR JAM

### Slimmer Fresh Berry Jam

2 tablespoons lemon juice
2 tablespoons water
1 envelope unflavored gelatin
3 cups crushed fresh strawberries or other berries
Sugar substitute equal to 1 cup sugar

1. Combine lemon juice, water, and gelatin in a saucepan and heat until gelatin is dissolved.
2. Add strawberries. Continue to heat the mixture rapidly to boiling.
3. When mixture boils, reduce heat to simmering; simmer for 2 minutes.
4. Remove from heat and stir in sugar substitute.
5. Ladle into hot, clean jars; cover, cool, and store in refrigerator.

Yield: 2½ cups

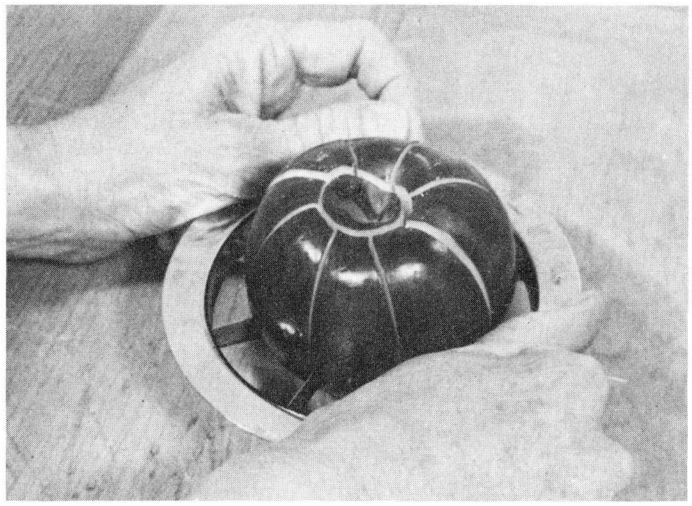

*Figure 10.32. Apples may be quartered for apple butter, or they may be sliced efficiently with the aid of an apple or pear slicer.*

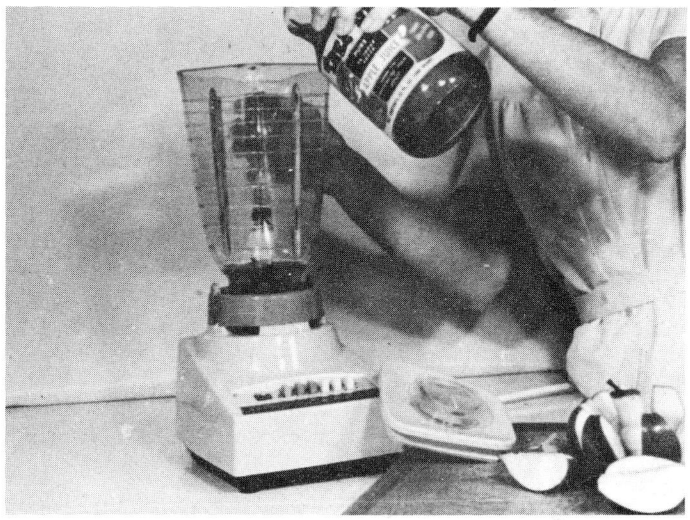

*Figure 10.33. A small amount of apple juice is used to facilitate the pureeing of the raw apple in the blender.*

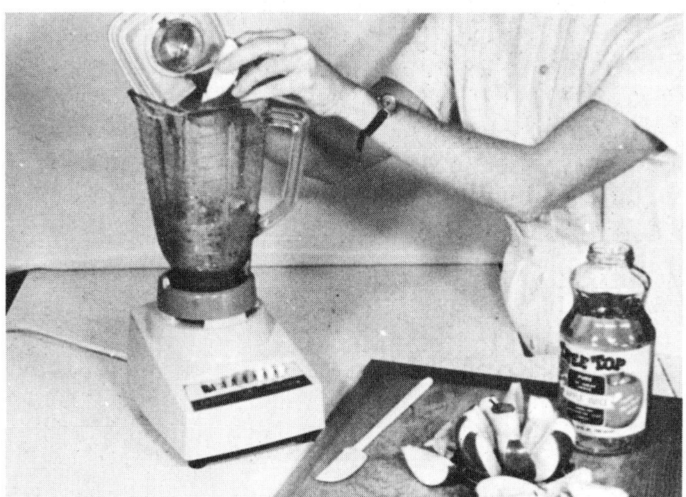

*Figure 10.34. More uniform size of pieces in the puree is accomplished by blending only a small amount of apple at a time.*

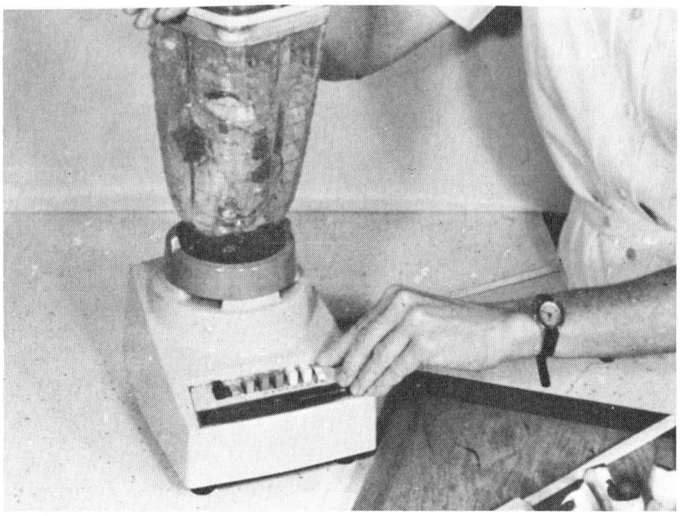

*Figure 10.35. The apple and apple juice are blended until two-cup quantities are prepared. These amounts are transferred to a wide, shallow pan, such as a roaster, until 16 cups of puree have been prepared.*

*Figure 10.36. The sugar and vinegar are stirred into the puree prior to baking.*

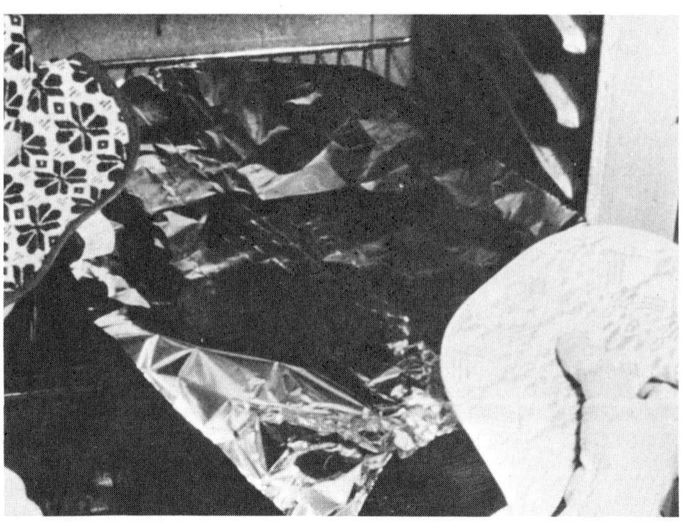

*Figure 10.37. Although the baking of the apple butter is done at the relatively low temperature of 325°F., the puree will splatter. Splattering is controlled by placing a piece of aluminum foil across the top of the pan loosely to permit evaporation to occur.*

*Figure 10.38. The spices are mixed together while the butter is being baked for the initial two-hour baking period.*

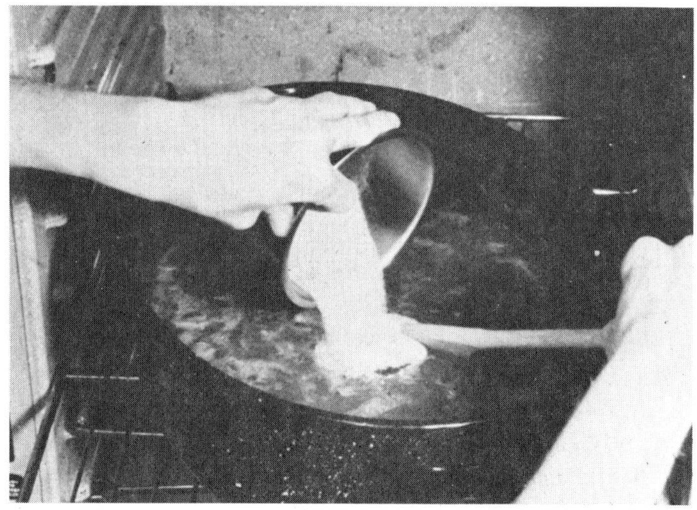

*Figure 10.39. At the end of the two hours, the spice mixture is stirred into the hot butter, the aluminum foil is again laid across the top of the roaster pan, and baking is resumed for another hour.*

*Figure 10.40. When the butter is uncovered after the third hour of baking, the volume will have been reduced perceptibly due to evaporation, and the color of the butter will be a warm, earthy brown-rust tone.*

*Figure 10.41. A spoonful of the butter is tested on a plate to determine the viscosity. When done, the butter will pile; no rim of liquid will form.*

*Figure 10.42. Since apple butter requires water bath canning for safe shelf storage, regular canning jars and lids are needed.*

*Figure 10.43. Apple butter is processed in a water bath canner for ten minutes after the water reaches an active boil.*

*Figure 10.44. Commercial gelatin is a protein product which can be used to form a gel structure. Flavored gelatins, such as the one illustrated, will include sugar as well as protein. Unflavored gelatin is a useful product for persons desiring to have jams or jellies with a reduced sugar content.*

## PREPARATION OF NON-PECTIN PRODUCTS

### Ripe Tomato Preserves

7 cups small pieces of peeled, ripe tomatoes
4 cups sugar
1 package (3 oz.) lemon-flavored gelatin

1. Bring tomatoes to a boil in a large kettle and simmer 10 minutes, stirring frequently.
2. Add the sugar and heat again to boiling.
3. Boil for 10 minutes while stirring continuously.
4. Remove from heat and stir in the gelatin until completely dissolved.
5. Immediately ladle into hot, clean jars or glasses and seal.
6. When cool, store in the refrigerator.

*Note:* The gel structure in this type of preserves is provided by the gelatin, which is a protein. Because of this difference, this type of preserves must be stored in the refrigerator, and for a maximum time of 1 month.

### Ever-Ready Red Fruit Jam

5 cups cut rhubarb (5 to 6 stalks)
1 package (10 oz.) frozen strawberries, raspberries, or cherries packed in sugar syrup
2 cups sugar
1 package (3 oz.) gelatin (same flavor as fruit used)

1. Combine rhubarb, thawed fruit, and sugar in a large saucepan.
2. Heat over low heat while stirring to dissolve the sugar. Increase rate of heating and quickly bring to a boiling.

3. Simmer just below boiling for 10 minutes, stirring constantly.
4. Remove from heat, add dry gelatin, and stir until dissolved.
5. Ladle jam into hot, clean jars; cover, and store cooled jam in the refrigerator.

*Note:* The gel structure in this type of preserves is provided by the gelatin, which is a protein. Because of differences in keeping qualities of protein-containing media and non-protein-containing jam, storage should be in the refrigerator for a maximum of 1 month.

# Chapter 11. Preservation by Freezing

## MICROBIOLOGICAL CONSIDERATIONS

Freezing is an effective method of preserving foods, for the low temperatures involved in storage are effective in curbing growth of microorganisms. However, the microorganisms which may be present in the food at the time of freezing have the potential for resuming growth if the temperature is allowed to rise at any time. Thus, safety of frozen foods is dependent upon the control of sanitation and the achievement of very low levels of microorganisms prior to freezing, a very fast rate of freezing, and well controlled frozen storage at consistently low temperatures.

### Microbiological Contaminants

#### Staphylococcal hazards

Gastrointestinal distress within approximately six hours after eating is an indication of possible staphylococcal contamination in frozen foods. The most likely foods in which this type of microbiological contamination may be found are moist, low acid foods. These bacteria are killed readily when subjected to heating, but their toxins are more resistant to heat treatment. However, with proper attention to sanitation during the preparation for freezing and good temperature control during frozen storage, these bacteria will not present a hazard.

#### Botulism

Ordinarily, frozen foods would not be a source of the toxin produced by *Clostridium botulinum*, but poor handling methods following freezing have the potential for creating this hazard. These bacteria may be present in the fresh food as it comes from the field or garden. Such contamination ordinarily is not a problem because the bacteria are destroyed by heat during preparation for freezing. The difficulty arises if spores are present and if the food is mishandled following frozen storage and prior to consumption. The spores are not destroyed by the temperatures ordinarily used in preparing for frozen storage. This does not present a hazard if foods are heated and consumed promptly after frozen storage. The problem develops when a food contaminated with spores is heated and then held many hours before being consumed. Such handling is hazardous because the spores remain viable during frozen storage, and toxin will be produced in the heated food if it is held for a long period of time prior to consumption. With ordinary precautions for food safety, this type of food poisoning will not occur in frozen foods.

#### Parasites

Foods have the potential for containing parasites, although the present controls exercised in food production have done much to reduce the likelihood that parasites actually will be present. *Cysticercus bovis* (larval stage of beef tapeworm), *Endamoeba histolytica* (amoebic dysentery), and *Trichinella spiralis* (trichinosis) are examples of parasites which can be found in food, although their occurrence is certainly rare. Freezing affords a very effective means of insuring that these parasites are not viable in foods. Frozen storage for at least a day at 0°F (-18°C) is effective in killing these organisms. Since most frozen foods are stored for longer than this, freezing represents a suitable means of eliminating this hazard in foods.

*Figure 11.1. Frozen storage of pork is effective as a means of killing* Trichinella spiralis *which could be present.*

### Other possible contaminants

Various other microbiological contaminants also may be found in frozen foods. Salmonella present prior to freezing may be killed by heat processing before freezing, but any viable organisms have the potential to resume their growth upon thawing of the food. *Shigella, Escherichia coli,* and *Bacillus subtilis* are examples of microorganisms which can cause food-borne illness if handled improperly following frozen storage.

## Survival of Microorganisms

### Frozen storage

Survival rates of microorganisms during frozen storage of foods are influenced by the type of microorganism and the pH of the food in which the contaminant is found. Foods which are relatively acidic, such as fruits, are less likely to have viable microorganisms following frozen storage than are low acid foods, such as vegetables. Survival rates in vegetables or other foods with minimal acid content may range from 1 to 75 percent, whereas the survival rate of harmless microorganisms in acidic foods has been found to range from 0.1 to 10 percent. In contrast, bacterial spores are far more resistant to destruction during frozen storage. Survival may be as high as 100 percent. Since it is clear that survival of viable microorganisms is to be expected, the handling of foods following frozen storage must be based on sound principles of food sanitation.

The actual temperature of frozen storage also influences survival of microorganisms. Temperatures slightly below freezing are more destructive than lower temperatures. *E. coli* destruction was determined to be far greater at -2°C than at -20°C. Similar results have been observed in a variety of foods, including beef and various vegetables. Even when liquid nitrogen has been used for freezing, this very low temperature has been less destructive than temperatures just below freezing.

### Minimum temperatures

Successful temperatures for frozen storage require that microorganisms be unable to multiply during storage. Unfortunately, the temperature range for growth varies with the specific microorganism. The microorganisms commonly causing food-borne illnesses are classified as mesophiles because they reproduce rapidly at moderate temperatures (within the approximate range of 40°F to 120°F). The microorganisms of particular concern during frozen storage are those which are capable of reproducing at the storage temperatures used for frozen foods. These are termed psychrophiles.

Psychrophiles are capable of growing not only at room temperatures, but also at temperatures somewhat below freezing. Psychrophiles found in foods include: *Pseudomonas* species, *Achromobacter, Flavobacterium, Escherichia, Aerobacter,* and *Alcaligenes.* Some molds also are classified as psychrophiles. *Penicillium* is an example of a psychrophilic mold.

Psychrophiles may prove to be detrimental to the quality of a food which has undergone frozen storage, but they do not present the hazard of food-borne illnesses stemming from frozen storage. Deterioration of color, texture, and general palatability may result from the action of psychrophiles during storage unless low storage temperatures are maintained. Psychrophiles may grow at temperatures as low as 10°F (-12°C), but they are not able to grow at temperatures colder than this. Consequently, if storage is maintained below this temperature range (as indeed it ordinarily is), psychrophiles will present no problem in the deterioration of food quality during frozen storage.

In foods which have been frozen and allowed to thaw, there will begin to be some competition between the mesophilic microorganisms and the psychrophiles as the food approaches room temperature. However, the mesophiles which could cause food poisoning may be suppressed in their growth by the spoilage microorganisms, i.e., the psychrophiles. The detrimental effects of the spoilage microorganisms likely would cause the food to be discarded rather than consumed. Thus, the psychrophiles serve as an aid in preventing food poisoning when viewed from the perspective of their role in influencing consumption.

Thawing of a food and subsequent refreezing can be done without health risk if none of the food reaches a temperature above 37°F (3°C) for microorganism growth is very slow up to this point. For practical purposes, this means that a food which has thawed on the surface, but which still has many ice crystals in the center can be refrozen. However, the quality of the food may be impaired by the refreezing. This possible loss of quality may be caused by growth of psychrophiles because of the food being held for a period of time in their favorable temperature range and also because of the possible cellular damage due to the reforming ice crystals.

## CHEMICAL AND PHYSICAL CONSIDERATIONS

A number of changes, both chemical and physical, may take place during freezing and subsequent frozen storage. Enzymes may catalyze chemical changes detrimental to the quality of the frozen food. The action of lipase is more likely to be a problem than is the activity of proteinases during frozen storage, but either type of enzyme may present quality problems. The treatment of the food prior to storage influences the extent of enzyme activity subsequently.

Physical changes also can be noted in foods which have been frozen. The clumping of egg yolk as a result of freezing illustrates this type of problem. Starch products also reflect physical changes viewed as undesirable.

### Changes in Flavor

Flavor, i.e., the combined impressions of odor and taste, may be modified during frozen storage as a result of enzymatic or oxidative changes in the food. Vegetables and foods containing fats are the types of foods most susceptible to detrimental changes in flavor during frozen storage. The extent of flavor modification is determined by treatment prior to freezing and by storage conditions while frozen.

Vegetables need to be blanched prior to being frozen to aid in minimizing flavor changes. The blanching procedure is effective in inactivating lipoxidase, the enzyme causing the development of rancidity in fats in vegetables. However, vegetables may have undesirable flavor changes during frozen storage even when blanching has been done. The chemical nature of these reactions leading to poorer flavors is not known at this time. Workers in the commercial frozen food industry have noted that bruised vegetables may develop a "hay-like" off-flavor if these products are held for a storage period prior to processing. Apparently, bruised vegetables have an alteration in their process of respiration, with the result that there is a decrease in oxygen consumption and an increase in carbon dioxide production. This difference may cause the off-flavor noted in bruised vegetables. For optimum quality of food frozen in the home, the food should be handled carefully and processed as quickly as possible to avoid the opportunity for this modified respiration pattern to develop.

The problem of off-flavors is greater in meats and other fat-containing foods than is the difficulty noted in vegetables. There are two types of rancidity which can develop: (1) hydrolytic, and (2) oxidative. Hydrolytic rancidity is the process which occurs when lipase catalyzes the cleavage of fatty acids from the fat molecule, resulting in free fatty acids and glycerol as the end products of the hydrolytic breakdown. Since this type of enzyme is inactivated readily by heat, this type of rancidity need not be a problem in frozen foods. Even the relatively mild heat treatment involved in pasteurization of milk is sufficient to inactivate lipase. Consequently, milk-containing products and foods which have undergone a heat treatment (such as blanched vegetables) are not likely to develop a rancid flavor due to hydrolytic action of lipase.

The problem of oxidative rancidity of fats is more complex. This type of rancidity involves the addition of oxygen to unsaturated fatty acids in fats, a chemical change resulting first in the formation of peroxides and leading to the production of many different compounds. This type of change occurs slowly at first; this period of slow change is the induction period. Once the free radicals begin to form, the reactions become autocatalytic, and the process accelerates, to the detriment of the flavor of the food.

Oxidative rancidity may include in its products aldehydes which subsequently are capable of reacting with proteins in a reaction termed the "Maillard reaction". This reaction causes a browning in the color of foods in which the reaction is taking place, a change which obviously is detrimental when occurring in light-colored foods, such as fish. However, from a flavor standpoint, the Maillard reaction is detrimental because it is a part of the total impression of rancid flavor developed in the complex reactions related to oxidative rancidity.

Various factors influence the likelihood of having flavor problems develop in frozen foods as a result of oxidative rancidity. The presence of metals, particularly iron and copper, may accelerate the development of oxidative rancidity in fat-containing foods. Research (Watts and Peng, 1947) indicated that precooking of meats can be effective in retarding oxidative rancidity, presumably because the iron bound in the proteins of meats is rendered less available as a result of heat denaturation of the proteins. Another factor contributing to the acceleration of oxidative rancidity is the presence of freezer burn. Apparently freezer burn accelerates oxidative rancidity by leaving the dehydrated surface more exposed to oxygen.

Oxidative rancidity can be impeded by the presence of antioxidants. The tocopherols are natural antioxidants which are effective in delaying the onset of oxidative rancidity. It is possible, but not economically feasible at the present time to modify the tocopherol levels in poultry by feeding high levels of the tocopherols, thus promoting greater resistance to oxidative rancidity when these meats are frozen. A more practical approach is to add antioxidants to fat-containing foods where the nature of the food makes such additions possible. The addition might be in the form of glazing the exterior surface of fish with an ice containing an antioxidant, or it might be incorporating of butylated hydroxytoluene (BHT), butylated hydroxyanisole (BHA), or tocopherols added directly to the fats used in formulating some foods.

Yet another approach to minimizing flavor problems due to oxidative rancidity is to eliminate as much air as possible from the package in which the food is frozen. Air-tight packaging is helpful because of reducing the oxygen content of the package. It also is valuable because it prevents freezer burn as long as the plastic is unbroken.

## Texture Considerations

The texture of frozen foods is modified by the freezing process; whether the change is viewed to be of significance is related to the intended use of the food following storage. For example, the crystal structure formed during freezing is of significance in determining the textural quality of ice cream because the crystals will be consumed. The formation of crystals during freezing of a fruit juice will be of much less concern since it is the thawed juice which ordinarily will be used.

Textural changes need to be reviewed in relation to modifications which may be caused by the effect of freezing on carbohydrates and proteins, as well as the physical damage to cells.

### Carbohydrates

Starch is a component of many foods which are frozen. During the preparation of starch-containing foods for freezing, the starch commonly is gelatinized. In the process of gelatinization, the starch granules absorb water into their structure, hydrogen binding water in the granule as the granule swells. At the same time, some of the amylose fraction in the starch granule is dissolved into the aqueous medium surrounding the granules. It is this dissolved amylose fraction which is of concern when viewing the possible textural changes in frozen, starch-containing foods. The amylose molecules can align themselves in frozen foods into a somewhat organized crystalline formation. Aggregates of these amylose crystals create a somewhat gritty texture when they are consumed. If a starch product is served without being heated after frozen storage, as would be true for a cream pie, the texture will be rough because of the amylose aggregates. This is a negative textural change which makes starch products that will be served cold poorly suited to frozen storage. However, starch-thickened foods which will be reheated after frozen storage and consumed hot have a pleasing consistency because the amylose crystals are dissolved when the product is heated.

The process in which amylose forms crystals in a gelatinized starch mixture is termed *retrogradation.* The process is reversed simply by heating the mixture because the hydrogen bonds which formed when the amylose crystals were established are broken by the heat energy, and the amylose once again is able to move in the mixture. Special starches, essentially pure amylopectin fraction, have been developed and are available for commercial use. These do not present the problem of retrogradation since they do not contain amylose. However, the consumer market does not generally have access to these starches. The most practical approach to solving the problem of retrogradation when freezing starch-thickened products to be served cold is to use rice starch or rice flour. These products retrograde less than do those which are thickened with flour (wheat starch) or with cornstarch, but the rice-thickened products still have a tendency to develop a coarse texture.

The choice of a suitable starch for thickening products to be served warm also needs to be considered. The retrogradation of wheat- and corn-thickened products causes a textural change, which can be detected by the eye as well as the mouth. Even when a sauce has been reheated, a curdled appearance can be seen in wheat-thickened and cornstarch-thickened sauces unless they are stirred quite vigorously. A sauce thickened with these cereal starches can be very satisfactory after reheating because it can be stirred. However, a meat pie filling or a casserole containing a white sauce cannot be stirred and, therefore, will maintain its unattractive, curdled appearance even though the textural problem is overcome by reheating. Rice starch does not produce sauces which have a curdled appearance when reheated; this difference makes rice flour or rice starch a better choice than either flour or cornstarch when preparing casseroles for freezing.

Glycogen is another polysaccharide (complex carbohydrate) which may be of interest in considering changes during freezing. In the freshly slaughtered meats, glycogen will undergo a breakdown to lactic acid as the carcass undergoes rigor. The formation of lactic acid reduces the pH of the flesh, a change which increases the tenderness of the meat. Meat to be frozen should be permitted to undergo the complete process of rigor rather than being frozen while the flesh is still in rigor. By allowing this

*Figure 11.2. This type of casserole is well-suited to freezing, but the ingredients will need to be stirred together after the dish has been thawed and heated so that the starch-thickened sauce will not have a curdled appearance.*

normal process to proceed to completion, the meat will achieve its maximum tenderness.

The effect of sugars on texture is related to the type of sugar and the quantity. High sugar content in ice creams has a detrimental affect on texture because of the limited solubility of sugars at low temperatures. Since only a small amount of sugar can be dissolved, any excess sugar in the mixture precipitates as crystals, contributing to the granular characteristic of the frozen product. The problem is complicated by the fact that milk contains lactose, and lactose is the least soluble of the disaccharides.

**Proteins**

One of the possible changes which proteins may undergo as a result of freezing is denaturation. When denaturation occurs as a result of frozen storage, the protein will become less soluble, as can be noted in some fish. Myosin, a protein found in cod, will denature during freezing and defrosting, with the result that the fish becomes tougher. This behavior is in contrast to the myosin in beef and poultry; myosin in these foods does not denature as a result of the freezing process.

Another example of a change in protein as a result of freezing is found in egg yolks. Apparently, the lipid-protein complex found in the yolk dissociates, and the protein precipitates, resulting in a rather pasty product. The addition of sugar or salt reduces this problem, seemingly by modifying the tendency of the protein to precipitate.

**Cellular disruption**

The texture of foods, particularly the texture of plant foods and meats, is influenced by the nature of the cell wall and by the fluid content within the cell. The structure of the cell walls of some fruits and vegetables clearly is very delicate. The texture of a food such as strawberries is dependent upon retaining a high fluid level within the cell so that the cell wall is extended fully. If the cell wall is ruptured by ice crystals during frozen storage or during the freezing process, the liquid will no longer be trapped within the cell, and the texture will become mushy. This loss of firm, crisp texture in fruits has long presented a problem in frozen food technology, a problem which still remains to be solved. Two approaches to moderating this problem can be used, but only with limited success. The first approach is to freeze such foods as rapidly as possible so that ice crystals will be of absolutely minimum size, thus causing less damage to the integrity of the cell wall. The most rapid method of freezing available is immersion in liquid nitrogen, which does produce a product with better texture than is achieved with slower rates of freezing. This method has potential for commercial freezing ventures, but is not applicable in the home freezing of foods. The other approach is to serve these delicate types of foods while they still have some ice crystals in them to provide structural interest rather than waiting until the food has thawed completely. This technique is applicable in the home. However, even when berries are served with some ice crystals remaining, they lack the perky appearance and interesting texture found in the fresh berries.

## Color Changes

Pigment changes can be noted in frozen beef. The color of beef is imparted by myoglobin, an iron-containing pigment which chemically is related to chlorophyll, and which, like chlorophyll, can undergo discernible changes in hue as a result of chemical change. During frozen storage, the iron in the compound can undergo an oxidative reaction to form metmyoglobin, a reddish brown pigment with less visual appeal than the original myoglobin or oxymyoglobin.

Some of the reddish colored fruits and vegetables become less attractive during frozen storage because of loss of some of their anthocyanin pigments into the medium surrounding them. Anthocyanins are contained in the vacuole; damage to the cell wall releases these water-soluble anthocyanins into the juices rather than retaining them in the cells. The result is a brilliantly-colored juice, but rather drab fruit. This phenomenon is seen in raspberries and sour cherries which have been frozen and stored for a fairly long storage period, particularly if the storage has been at temperatures just below the freezing point.

Discoloration of the bones of frozen poultry presents an aesthetic problem, simply because one does not expect to see dark bones. The dark color often observed in the bones of frozen chicken which has been cooked is due to the breakdown of the red blood cells in bone marrow during freezing and the movement of the freed hemoglobin into the bone structure. Upon cooking, this displaced hemoglobin denatures and gives the bones the dark appearance so often observed in cooked frozen and thawed poultry.

Enzymatic action can produce color changes in some frozen foods. A notable example is provided by peaches. Peaches contain an enzyme, polyphenoloxidase, which catalyzes a chemical change of the flavonoid pigments in peaches. Polyphenoloxidase catalyzes the reaction of molecular oxygen with the phenolic pigment to produce quinones, which in turn are converted into brown-colored polymers, yielding the familiar brownish color of frozen peaches. The presence of an antioxidant, such as ascorbic acid, is effective in blocking the reaction until the antioxidant supply has been exhausted. This ability of antioxidants to react with the oxygen found in the package protects the pigments in the peaches. This is the reason for including ascorbic acid as an ingredient when preparing frozen peaches. It is apparent that the polyphenoloxidase is not an enzyme which is inactivated by freezing. The only approach presently known to counteracting its influence is to utilize an antioxidant.

## NUTRITIONAL CONSIDERATIONS

The relative nutritional merits of frozen foods are influenced by the preparation required prior to freezing. The blanching of vegetables is necessary to inactivate enzymes and enhance retention of quality during frozen storage, but the blanching process is somewhat destructive to the content of water-soluble vitamins. Loss of vitamin C as a result of blanching may run as high as 40 percent. The B vitamins are less susceptible to destruction than is vitamin C, but losses of the B vitamins also are engendered as a result of blanching.

The rate of freezing, whether foods are frozen in liquid nitrogen or by other far slower means, appears to have little influence on the retention of vitamins. From the standpoint of home freezing, this is helpful because extremely rapid rates of freezing cannot be achieved with present home equipment.

The thawing of frozen fruits and vegetables appears to have little influence on the vitamin content, but this is not true for meat and poultry. The loss of the water-soluble vitamins into the drip of thawing meats is related to the amount of cut surface, ranging from as little as 3 percent in whole turkeys to 10 percent in steaks.

## EQUIPMENT FOR FREEZING

One of the advantages of freezing as a method of food preservation is that special equipment is not required. The freezing of vegetables requires some means of blanching the vegetables prior to freezing. This may be as simple as using a large piece of cheesecloth, or it might mean a large kettle with a basket to hold the vegetable while immersing in the boiling water, or it could be a steamer. Other types of foods to be frozen do not require even this much preparation equipment.

The containers for frozen foods can be of plastic or glass if liquid materials are to be frozen. Solids, such as baked products and meats, may be packaged in an air-tight wrap rather than in a shaped container. The important consideration in packaging for frozen storage is that the food must be protected from surface drying in the freezer while it is being stored.

Freezing requires adequate storage space where the temperature can be maintained at 0°F or lower. The freezer compartment in a refrigerator may be used for short periods of frozen storage, but for storage periods anticipated to range from 3 months to a year, a home freezer is a more appropriate device.

## MAINTAINING THE FREEZER

All frozen foods need to be labeled clearly with the name of the food and the date of freezing clearly identified so that they can be found quickly in the freezer.

If much food is being held in the freezer, a chart indicating the location of the various types of food and their date of freezing will be a valuable aid in finding foods. This approach to frozen storage is important because the freezer door should be held open as short a time as possible when removing foods. This measure helps to avoid undesirable variations in freezer temperature.

The chart, including freezing dates, is helpful not only in finding the desired food, but also in maintaining an inventory of the foods stored in the freezer. Optimum storage times vary with the type of food which is being stored. Ice cream and other frozen desserts are best when used within a month of freezing because of their tendency to develop coarser crystals. At the other end of the storage spectrum are products such as fruits, which can be stored satisfactorily for up to a year.

## Defrosting

For optimal storage in a freezer, it is necessary to avoid building up heavy layers of frost on the walls of the freezer. This precaution will enable the freezer to operate most efficiently and to maintain appropriately low storage temperatures. The time between major defrostings can be extended by scraping off frost as it begins to build up in the freezer. A plastic scraper or spatula is an effective tool for this job. Only a very brief time should be used for this operation. Otherwise, the air from the room will simply add to the moisture available for forming frost in the freezer.

Even when occasional brief scraping has been done and when the door has been opened infrequently, it still will be necessary to defrost the freezer occasionally. This can be done best by reducing the temperature of the freezer overnight to its lowest setting so that the food will be as cold as possible before the defrosting operation is begun. The actual defrosting should be done as quickly as possible. The freezer is disconnected and the food quickly removed from the freezer. By placing pans of hot water in the compartment and scraping carefully, but quickly with the plastic scraper, the removal of the ice can be speeded considerably. All of the ice needs to be removed and the entire interior wiped dry before the freezer is turned on again. The control should be on the lowest setting while the freezer begins to cool its compartment. After about ten minutes of operation, the food can be replaced quickly in the compartment. Ideally, the defrosting is done at a time when the freezer inventory is low. This will enable the freezer to quickly cool the food again and resume its normal operation. When the food has been thoroughly chilled again to 0°F, the freezer control can once again be set to its normal operating position.

## Emergency Procedures

Sometimes there may be a power failure. This can create a significant hazard to frozen foods if it is not detected and rectified. Before undertaking a particular plan, it is wise to determine how long the problem will continue. Dry ice can be placed in the freezer to help maintain the necessary low temperatures if it appears that the food will thaw before power is restored. It can be anticipated that a fully loaded freezer, if kept closed during the power outage, will maintain food safely for up to two days, while a half load may stay frozen for only a day. If it appears that dry ice will be needed, 50 pounds of dry ice will maintain the food in a loaded freezer safely for up to four days; a partially-loaded freezer will be kept safe for two to three days by this amount of dry ice. The dry ice should be placed on pieces of cardboard positioned on top of the frozen goods and the freezer door closed again as quickly as possible. The dry ice should be handled with tongs, hot pads, or other protection; it should not be handled with the hands because it will injure the skin seriously. It also is important that there be adequate ventilation in the room while the dry ice is being transferred to the freezer because there is considerable carbon dioxide sublimating from the solid carbon dioxide of the dry ice to form the gaseous carbon dioxide at room temperature. The dry ice storage provides a solution when there is a power failure, but for mechanical problems in the freezer, it may be necessary to attempt to utilize another freezer until the one that failed can be repaired.

## Refreezing Foods

It is important to check regularly to be certain that the freezer is maintaining its expected low temperature of at least 0°F. This check will aid in averting loss from foods accidentally thawed by a malfunctioning freezer. If foods are found to be thawing, they may be used and even refrozen if ice crystals still are contained in the food. If foods have reached an interior temperature greater than 40°F, they represent a potential hazard and should be used only if they are still quite cold and have absolutely no indication of poor color or an off odor. Foods which are still cold, but which no longer contain ice crystals should not be refrozen. Foods which have been refrozen should be used very soon so that their quality will be as high as possible.

# PROCEDURES FOR FREEZING

## Fruits

### Selection

Many fruits freeze extremely well, although changes in their texture always will be noted. The color and flavor of rhubarb, cherries, berries, and pineapple will be excellent when frozen. The structure will be modified by freezing, but these fruits still may be used satisfactorily in many fruit desserts and sauces. Peaches and pears also may be frozen, but attention needs to be directed toward preventing discoloration. Melons can be frozen; the problem with melons of all types is the development of an almost rubbery texture following freezing and thawing. The texture change is considered by many people to be unacceptable. As a result, these fruits are frozen less frequently than others.

In selecting fruits for freezing, it is wise to decide whether the changes in texture which will occur inevitably will be acceptable to the consumer. If so, the selection process then involves only care in procuring fruits of high quality which are fully ripe, yet not too ripe and are free of blemishes. Such fruits will have maximum eye appeal following freezing and will also have excellent flavor.

**Preparation**

The first step in preparing to freeze fruits is to prepare the plastic containers or glass jars which are to be used for packaging. Rectangular plastic containers are excellent choices because they may be stacked and packed efficiently in the freezer where space often is at a premium. Plastic pouches which can be heat sealed provide another alternative for packaging. If glass jars are used, it is helpful to have freezer jars designed to be somewhat larger at the mouth than at the base. This design makes it easy to remove foods from the jars while still frozen. It also facilitates stacking in the freezer.

The containers for freezing need to be washed thoroughly with soapy water and then rinsed well. In contrast to the preparation of jars for jams and jellies, the jars used in freezing are allowed to cool after washing and rinsing. The closures also need to be washed and rinsed.

If a syrup pack is to be used, the syrup is prepared and chilled at this point. Then the fruit is prepared. All fruits are washed thoroughly in preparation for freezing. Any soft spots are cut out, the cores or pits are removed, and the fruits are sliced, mashed, or left whole as desired.

For best results in freezing at home, only small quantities (two or three quarts) should be done at one time. By using small amounts, and also by packing in half pints or pints, freezing of the fruit will occur more rapidly in the freezer. This will aid in obtaining optimum texture in the frozen fruit.

Pears, apples, peaches, and apricots require some protection if discoloration is to be avoided. These fruits can be enhanced by placing the freshly prepared fruit in a water and lemon juice solution (2 quarts of water and 2 tablespoons of lemon juice) briefly until the fruit is ready to be packed. Another method to prevent browning is to use ascorbic acid in the manner specified by the manufacturer. This may be by dissolving the vitamin in a small amount of water and adding to the sugar syrup which will be used to pack the fruit, by dissolving the vitamin in a little water and sprinkling it directly over the fruit if the fruit is being packed without sugar, or by sprinkling the vitamin (dissolved in a little cold water) on the fruit just before adding the dry sugar if this is the method of packing selected.

A few fruits, notably apples, apricots, blueberries, and rhubarb, need to be scalded briefly before being frozen. These fruits should be chilled in ice water to cool them completely and then drained thoroughly before packing for freezing.

**Packing**

Fruits may be packed in any of three ways: (1) dry sugar pack, (2) syrup, or (3) without sugar. As a rule of thumb, fruits intended to be used without being cooked in a recipe following freezing are best when frozen in a syrup. This would apply to fruits intended to be consumed as sauces or as toppings for cakes, puddings, and ice cream. For fruits in-

*Figure 11.3. Plastic containers with tight-fitting lids are excellent for storing whole fruits and purees or other fluid products. The design of the container permits efficient stacking and storage in the freezer after the food has been frozen.*

*Figure 11.4. Glass freezer jars also are suitable containers for storing frozen foods. The jars need to be designed without a shoulder so the food can slide out easily before it is thawed completely.*

*Figure 11.5. Apples and other fruits which tend to brown quickly, need to be treated with ascorbic acid or other acidic solution as soon as they are cut. The three slices on the left were not treated; the three on the right were dipped in an ascorbic acid solution.*

tended to be incorporated into recipes requiring cooking following frozen storage (for example, cherries for use in pies), a dry sugar pack or even packing without sugar is appropriate.

Syrups are made by bringing the measured amount of sugar and water to a boil and heating until the sugar dissolves, and the mixture becomes clear. The resulting syrup should be refrigerated until it is almost ice cold. This is important as a means of helping the fruit product to freeze as quickly as possible in the freezer. The concentration of the sugar syrup may be selected to meet personal preference and to offset the acidity of some fruits. Usually a 40 or 50 percent syrup is used for packing fruits in syrup, and a cup of syrup is allowed for each quart of fruit to be frozen. This is merely a guideline for preparing the syrup required for the amount of fruit to be processed at one time.

Fruits to be packed with the syrup pack are placed in the freezing container in which approximately half a cup of syrup has been placed. The container is filled gently with fruit until it is about half full. At this point, the fruit is shaken a bit to help pack it gently without any crushing. The container then is filled with more fruit until half an inch of headspace (3/4 inch for 1½ pints) remains. Again the jar is shaken to distribute the fruit, and more fruit is added if additional space is available. If the juice is not sufficient to cover the fruit, more syrup is added until the fruit is covered. A small piece of crumpled plastic wrap or aluminum foil is placed on top of the fruit before the cover is placed in position. This measure will help to keep the fruit from floating above the surface of the syrup. If any syrup or fruit has spilled on the rim of the container, this should be wiped away before the cover is put on.

Fruits to be packed in dry sugar are first placed in a shallow pan and sprinkled with dissolved ascorbic acid if it is needed to prevent browning. Then sugar in the recommended amount is sprinkled over the fruit. The fruit is stirred very gently to help coat all of it with the sugar and to aid in forming a juice which uniformly coats the fruit. The fruit then is placed in the container and gently shaken to aid in filling the jar, yet not causing the fruit to be crushed. The juice collecting from the fruit is spooned over the fruit into the container. There should be enough juice to cover the fruit. Plastic wrap or aluminum foil is placed on top to help keep the fruit under the juice. A headspace of the same amount as described in the syrup pack is recommended. The rims should be wiped before the container is closed.

Fruits packed without sugar are simply sprinkled with the dissolved ascorbic acid (if it is required) and then packed and shaken gently in the container, leaving a headspace of half an inch. After wiping the rim clean, the container is closed. Crumpled plastic wrap or foil can be placed on the food to minimize ice crystal formation on the food.

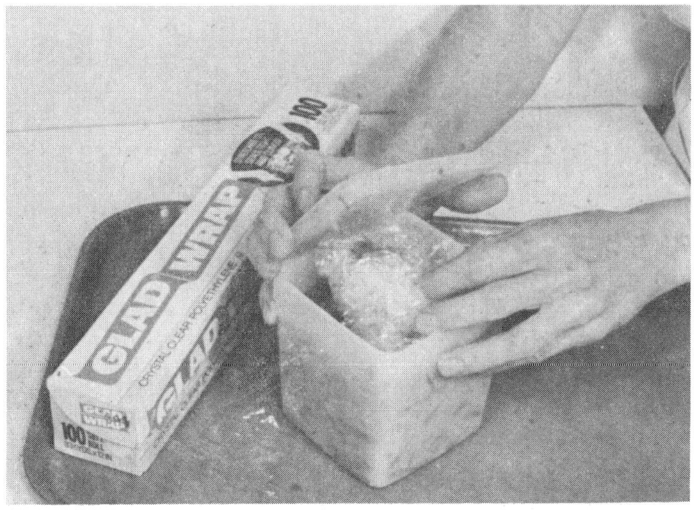


*Figure 11.6. Plastic wrap is crumpled and placed in the top of the container in which syrup-packed fruit is packed. The wrap holds the fruit beneath the surface of the syrup.*

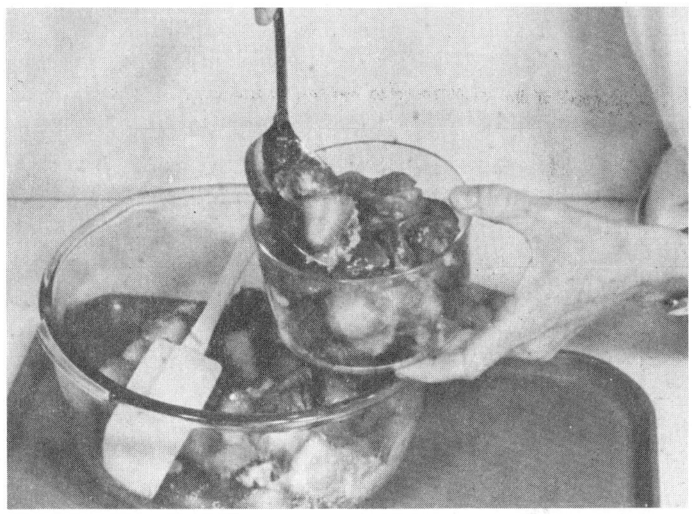

*Figure 11.7. A dry sugar pack is a quick method for packing fruits for freezing. Sugar is measured into the bowl with the prepared fruit and stirred gently to coat the fruit with the sugar. A headspace of approximately 3/4 inch is left at the top of the container.*

### Freezing and storing

As soon as the fruits are packed, they should be placed in the freezer in an upright position. For best results, they should be positioned so that there is space between each of the containers. This will insure that freezing takes place as rapidly as possible. When frozen, the containers may be stacked or stored more compactly in the freezer. All containers need to be labeled clearly with their contents, the amount, and the date of freezing before being stored in the freezer.

## Vegetables

### Selection

Vegetables for freezing need to be at the peak of quality; they should be young and fresh, flavorful, and free of blemishes. Many different vegetables can be frozen with satisfactory results. Peas, green beans, corn, and greens fare particularly well during frozen storage, particularly if carefully blanched prior to freezing. Asparagus and winter squash are also quite excellent for freezing, although they may undergo more textural change than those previously mentioned. Detectable differences in texture are unavoidable in the freezing of carrots, broccoli, cauliflower, and beets. However, the frozen products from these vegetables are still of very satisfactory quality. The changes in these types of vegetables need to be weighed against the availability of the fresh vegetable at various times of year and the relative costs of the fresh and frozen food.

*Figure 11.8. For optimum quality in frozen vegetables, freezing should be done immediately after harvesting.*

### Preparation

The preparation of the containers for freezing vegetables is the same as that described for freezing fruits. However, the preparation of vegetables for freezing requires an additional step because blanching is required to halt enzyme action, particularly the catalase and peroxidase found in vegetables such as cauliflower, spinach, and peas, as well as in most other vegetables. In preparation for blanching, the vegetables should be washed well, trimmed according to the specific requirements of the vegetable, and cut as desired. Only two or three quarts of vegetable should be processed at a time so that freezing can be completed very quickly in the freezer. While the vegetables are being prepared, the water should be heated for blanching.

Blanching is a necessary process in the freezing of all vegetables. This is done most easily and accurately by bringing a gallon of water to a rolling boil in a large kettle and then placing the prepared vegetable, no more than a quart at a time, in a metal basket or cheesecloth bag which is immediately lowered into the water. A lid is placed over the kettle to help bring the water back to a rolling boil very quickly. The blanching period is started at the time the vegetable is placed in the boiling water and is continued for the exact period of time recommended for the specific vegetable. Timing is critical; too long a blanching period creates undesirable textural changes, and too short a period will lead to the development of off-flavors and premature changes in color due to failure to inactivate the enzymes.

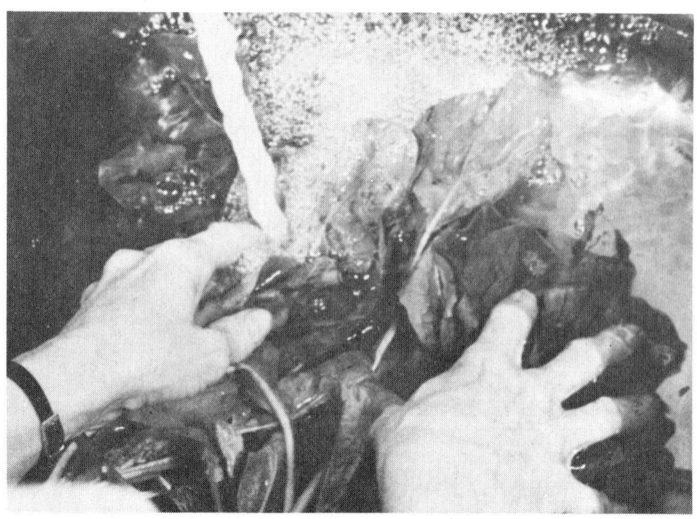

*Figure 11.9. Spinach and other greens freeze well, but it is imperative that very thorough washing be done prior to freezing to remove all dirt and sand which is trapped in the convolutions of the leaves.*

Steam blanching also can be done. Devices for steaming can be varied in design and material, but basically they consist of some type of rack to hold the vegetable in active steam above actively boiling water. Steaming requires about two more minutes of blanching time than is required in boiling water.

Allowance also needs to be made for temperature differences due to altitude. When blanching is done at an elevation of 2,500 feet above sea level, an additional blanching time of half a minute needs to be provided, and at 5,000 feet, the time should be extended by a minute if blanching is in boiling water and for a minute and a half if blanching is in steam.

*Figure 11.10. Blanching in boiling water is an important step in preparing vegetables for freezing. The heat reduces problems of deteriorative change during storage. By using a large kettle of water, the boiling water will boil very quickly again after the vegetable is added, thus reducing the length of time the vegetable is heated.*

*Figure 11.11. Steam blanching is a desirable method for preparing vegetables because there will be less loss of the water soluble vitamins during blanching.*

*Figure 11.12. Steam blanching requires that a lid be used to trap the steam in the pan to facilitate blanching of the vegetable suspended on the rack.*

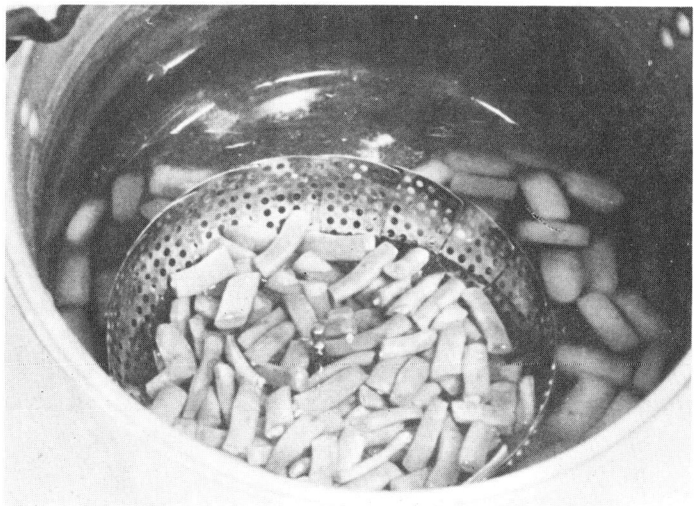

*Figure 11.13. As soon as the blanching period is completed, the vegetable immediately is immersed in ice water to halt the cooking action and cool the vegetable to hasten the freezing of the product.*

Just as soon as the blanching period is over, the vegetable should be placed in ice water to halt the cooking process immediately. Since no more than a quart of vegetable is blanched at a time, the hot vegetable will cool quickly. The time in ice water should be appproximately equivalent to the time in the boiling water. The vegetable then is drained thoroughly in preparation for packing.

### Packing

Vegetables, unlike fruits, are packed without any added liquid or juice. The blanched vegetables will fit efficiently in the containers, but they should not be packed tightly. The rims are wiped clean, and the containers are sealed. Note that the vegetables are packed without salt. If desired, crumpled plastic wrap can be placed between the vegetable and the cover to help minimize ice crystal formation on the surface of the product.

### Freezing and storing

When the containers have been closed, the packages quickly are labeled with the name of the product and the date of freezing.

The containers immediately are placed in the freezer, leaving ample room between containers to permit good circulation to facilitate freezing. After freezing has been accomplished, the containers can be consolidated more compactly in the freezer compartment, and their location noted on the freezer inventory.

## Meats

### Selection

Only meats, fish, or poultry of excellent quality should be selected for freezing. If the item to be frozen has been freshly slaughtered, it is necessary to let it pass through rigor before freezing is done. For these foods, it is imperative that all holding be done under very clean conditions and that the foods be chilled below a temperature of 40°F within 24 hours of slaughter. The holding time following this initial chilling is 1 to 2 days for pork and veal, between 5 and 7 days for lamb and beef, but chicken and other poultry can be frozen after a holding period of only 12 hours after the fowl has been placed in refrigerated storage. These chilling periods provide adequate safety measures while also promoting excellent tenderness.

The quantity to be selected for freezing is influenced significantly by the circumstances. However, it is wise to keep in mind the maximum storage period recommended for the type of food being frozen and to calculate how much realistically could be consumed within that time period. Sometimes considerable economies can be realized by prudent purchases of meat for freezing, but these economies may fail to materialize if more is frozen than can be eaten during the recommended period of storage.

Sometimes the opportunity to buy meat for the freezer begins with the live animal. If a decision must be made about such a purchase, it is helpful to estimate the approximate yield which will result from the carcass. Excluding the variety meats from beef, the edible cuts from a mature steer will equal approximately 45 percent of the live weight of the animal. The hindquarters represent approximately 45 percent of the trimmed cuts, and the forequarters constitute the remaining 55 percent of the edible yield. The weight of trimmed cuts from lamb is approximately 44 percent of the live weight, a figure very close to that from beef. The yield of pork in relation to live weight is significantly higher (55 percent). For a typical lamb carcass, a carcass weighing 41 pounds would yield approximately 31 pounds of legs, chops, and shoulder cuts, but only 7 pounds of breast and stew meat. From a 176 pound hog carcass, there would be approximately 90 pounds of hams, shoulders, bacon, and jowls; the remaining portions would yield about 34 pounds of loins, ribs, and sausage and 12 pounds of lard, when rendered. Because of the size of the animal, a fairly common practice is to buy either a hind- or a forequarter of beef, whereas some families might buy the whole lamb carcass. The cuts which result from cutting the forequarter and hindquarter are outlined in Table 11.1.

Table 11.1. Approximate yield from a 300 pound side of beef

| Forequarter | Pounds | Hindquarter | Pounds |
|---|---|---|---|
| Rib roast | 18.3 | Round steak | 27.0 |
| Blade chuck | 26.7 | Rump roast | 9.9 |
| Arm chuck | 17.4 | Tender steaks | 40.2 |
| Brisket | 6.3 | Flank steak | 1.5 |
| Lean trim | 49.2 | Lean trim | 21.0 |
| Waste (bone, fat) | 38.1 | Kidney | .9 |
| | | Waste (bone, fat) | 43.5 |
| Total | 156.0 | Total | 144.0 |

In addition to considering the total weight of meat which reasonably may be consumed within a reasonable period of frozen storage, careful thought needs to be given to the food preferences of a family. Unless all of the cuts from the carcass, side, or quarter will be used, purchases of these quantities may be false economies. Families with relatively narrow ranges of meat preferences may effect greater savings and more enjoyment by buying preferred cuts when they happen to be on special sales at the market.

**Preparation for freezing**

Preparation of meats for freezing may involve simply the cutting of the carcass, side, or quarter into the desired cuts. The thickness desired for the steaks, chops, or roasts should be specified in relation to the personal preferences and the cooking methods which likely will be used in preparing the meat after frozen storage. The cutting and handling of all meats need to be done under very sanitary, chilled conditions to avoid unnecessary contamination of the meat.

If poultry has been purchased dressed, the only preparation necessary is to check carefully for any feathers or lung tissue and to wash the carcass carefully inside and out with cold, running water. After being drained well and patted dry with paper toweling, the poultry is ready for packaging and freezing.

Poultry to be frozen should never be stuffed because of the microbiological hazard it presents. The stuffing in the cavity impedes the freezing process by presenting a far greater cross section to be frozen and by blocking the chilling which ordinarily occurs from the direction of the body cavity. Freezing temperature is reached far more quickly when the cross section of food is relatively small and the chilling is proceeding from both sides of the flesh.

Fish to be frozen often may be the result of a fishing expedition, which means a knowledge of cleaning and dressing fish may be required in preparation for freezing the catch. The first step after washing the fish is to scale it by scraping along the body of the fish with the dull edge of a knife, proceeding from the tail toward the head. The scales need to be rinsed off well under running water. Next, the head is cut off, the belly is split with a sharp knife the length of the fish, and the entrails are removed. The dorsal fin is removed by cutting through the flesh on both sides of the fin so that the bones at its base will lift out when the fin is pulled from the back of the fish. The fish then is scrubbed thoroughly both inside and out. At this point, the fish can be frozen or else cut into the desired portions. Steaks are cut by cutting cross-sectional slices of the fish between ¾ and 1 inch thick, perpendicular to the backbone. Fish fillets are made by cutting from the head to the tail of the fish along the backbone. This initial cut is followed by running the knife along the rib bones to remove the flesh from one side of the fish in a single piece, from the collarbone to the tail.

*Figure 11.14. Fish are scaled by scraping the dull edge of a knife gently from the tail toward the head.*

*Figure 11.15. Fish steaks are made simply by cutting crosswise through the cleaned fish in parallel cuts of the desired thickness. Usually steaks are cut an inch thick.*

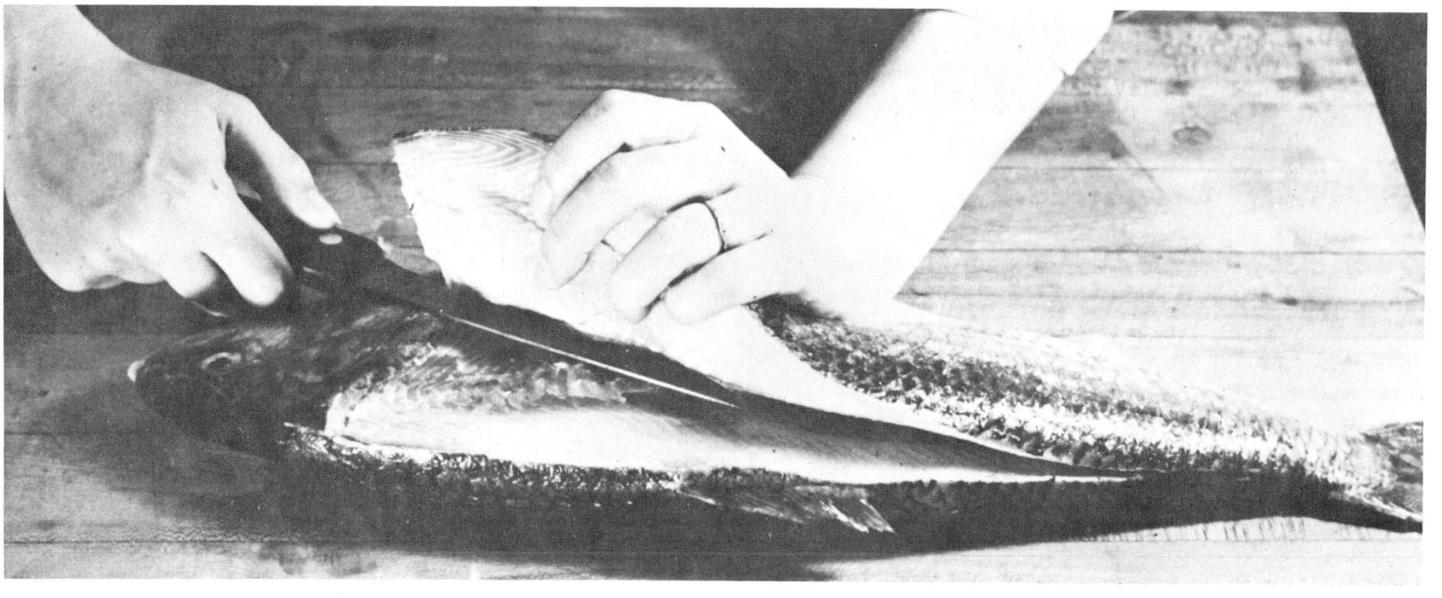

*Figure 11.16. Fillets are cut from the cleaned fish by first cutting down the back of the fish from head to tail. Then the flesh is removed by cutting the fillets free from the ribs and back bone.*

*Figure 11.17. When the flesh has been cut free from the bones, the fillet can be lifted out, but it needs to be handled gently to avoid tearing the flesh.*

After the first side has been filleted, the fish is turned over, and the second side is filleted in the same manner. If desired, the skin can be removed from the fillet before packaging for freezing. The skinning is done easily by placing the fillet with the skin side down on the cutting board, and then cutting with a sharp knife through the flesh to the skin, and then tipping the knife to run it directly along the skin to the other end of the fillet, freeing the flesh from the skin. The prepared steaks and fillets suitably are dipped in a 5 percent brine solution (2⅔ tablespoons of salt in 1 quart of water) for half a minute before being packaged for freezing.

Some kinds of fish have a dark layer of flesh extending the length of the back. To avoid the strong and somewhat unpleasant flavor this muscle imparts, it is suggested that this dark layer be removed from the fish prior to freezing.

### Packaging

Meats to be frozen should be packaged in the quantity desired for cooking at one time. Large packages of meat are very slow to freeze and slow to thaw. Therefore, relatively small packages have distinct advantages.

The important consideration in freezer packaging of meats is to use an air-tight packaging material so that the product will not develop freezer burn. Plastic freezer bags or heavy weight aluminum foil are suitable choices which generally are available. These materials can be wrapped tightly around the item to reduce the air content of the package to an absolute minimum to help reduce the problem of dehydration during storage and also to minimize the tendency toward the development of rancidity in the fat. Roasts are wrapped suitably by placing them in the center of a piece of heavy aluminum foil which is long enough to overlap itself by at least 3 inches when wrapped around the meat. The foil ends then can be brought together above the meat and folded down together to make a relatively air-tight seam. The two ends of the package are closed in a similar fashion after the foil has been pressed firmly against the surfaces of the meat to help eliminate air from the package.

Plastic freezer bags also may be used for storing meats, poultry, or fish in the freezer. As with the use of aluminum foil, it is necessary to force as much air as possible from the package prior to sealing tightly. A convenient means of removing much of the air is to insert a plastic straw in the package and suck the air out by inhaling.

Some plastic bags may be heat-sealed very satisfactorily; others need to be secured by twisting the opening very tightly after the bag has been evacuated and then taping or securing with a metal twister.

When steaks or fillets are being prepared for freezing, it may be more convenient to thaw them if they are packaged by placing layers of aluminum foil between the pieces or even by packaging them separately. For freezing poultry, the giblets are placed in aluminum foil or a plastic bag before being placed in the cavity of the fowl for packaging of the whole bird.

### Labeling and freezing

All packages of meat, poultry, or fish should be labeled very carefully with the name of the cuts, the number of cuts, the type of meat (or poultry or fish), and the date of freezing. When roasts are labeled (including whole fish to be baked or whole fowl for roasting), it is helpful to mark the weight on the package to serve as an aid in calculating roasting time when the roasting is done.

Turn the control in the freezer to its coldest setting before placing the meat in the freezer to aid in freezing the meat as quickly as possible. The control should be left at this setting for about 24 hours if the freezer is fairly full or for 8 to 12 hours for freezers half full. It is best to limit the quantity of food being added to the freezer at one time to two pounds per cubic foot of total freezer capacity. The packages of meat being placed in the freezer should be touching the walls or any other refrigerated portion of the unit. If being placed on top of other foods, a piece of cardboard should be laid between the two packages so that the unfrozen meat will not warm the frozen food.

### Thawing meats, poultry, and fish

Meats, fish, and poultry (particularly poultry frozen in pieces rather than whole) can be cooked while still frozen. This will require an extended cooking time, ordinarily approximately half again as long as would be needed if the meat had been thawed completely before being cooked. If the meat is frozen in smaller pieces rather than in a large roast, the added cooking time will be somewhat shorter than this, but still will be longer than if the meat were thawed first.

Thawing prior to cooking results in more uniform heat distribution in the meat during the normal time required for cooking. Thawing is best done by simply moving the package from the freezer to the refrigerator without even unwrapping it. The item also may be thawed by placing it in cold water if it has been bagged securely in a plastic bag. The water is changed frequently to facilitate the thawing process. Sometimes a combination of refrigerator and water bath thawing is appropriate when there is not time to thaw the item completely in the refrigerator. The refrigerator method is to be preferred when it can be used because it does not have the potential for microbiological hazards which are possible with thawing outside the refrigerator. Table 11.2 presents the anticipated thawing times for thawing whole poultry in the refrigerator.

## Freezing Other Foods

### Eggs and egg-containing dishes

Egg whites can be frozen very satisfactorily before they are cooked. Because of their excellent

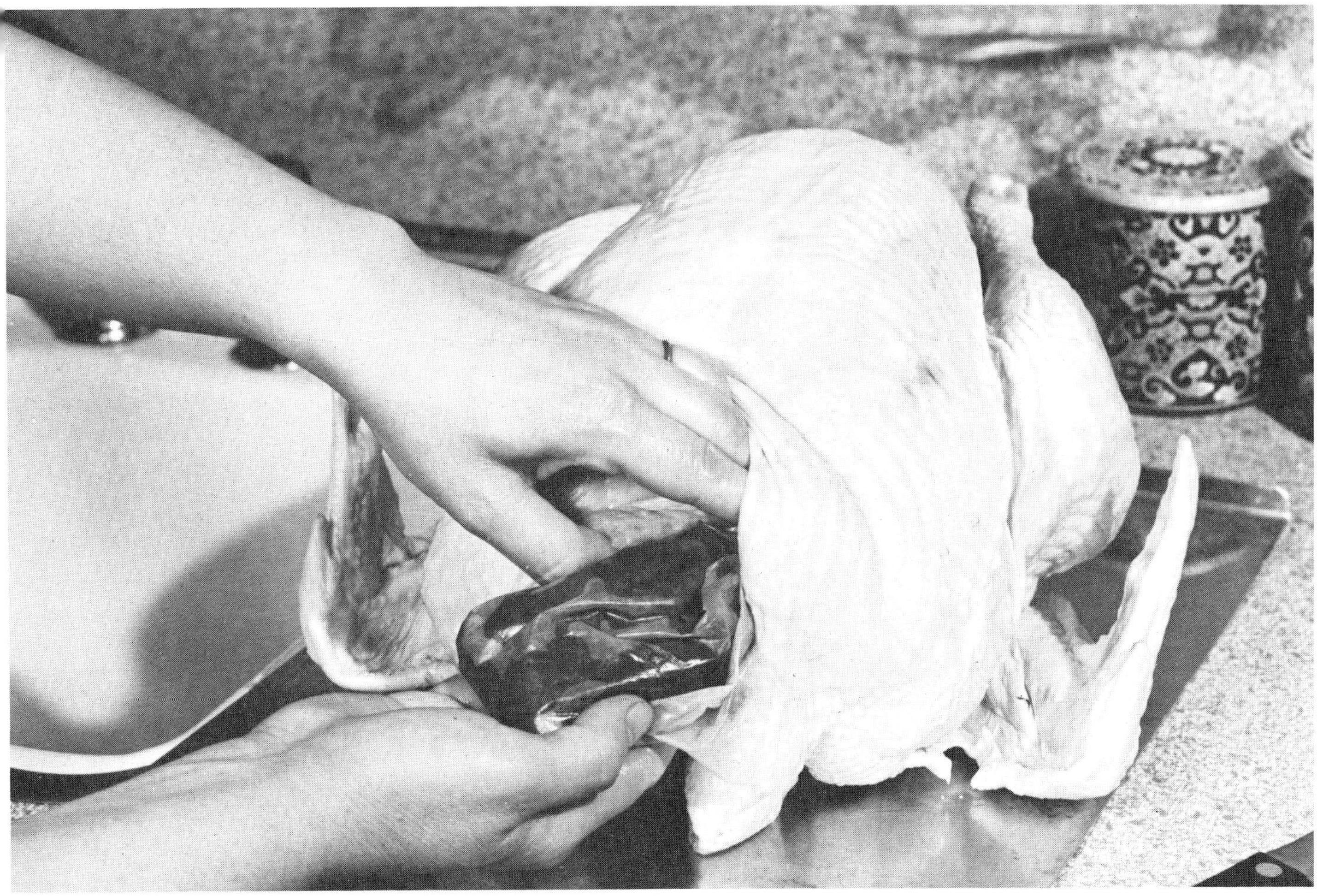

*Figure 11.18. The giblets should be packaged in a plastic bag before being placed in the cavity for freezing poultry.*

Table 11.2. Timetable for thawing whole poultry in the refrigerator

| Type of fowl | Time to thaw |
|---|---|
| Chickens: | |
| 4 pounds and over | 1 to 1½ days |
| Less than 4 pounds | 12 to 16 hours |
| Ducks, 3 to 5 pounds | 1 to 1½ days |
| Geese, 4 to 14 pounds | 1 to 2 days |
| Turkeys: | |
| 18 pounds and over | 2 to 3 days |
| Less than 18 pounds | 1 to 2 days |

performance in foams following thawing, frozen egg whites present an excellent means of preserving whites which may be left when yolks are the only portion of the egg required in a recipe. Freezing of the whites is done simply by placing the whites in a container, covering tightly to prevent evaporation, and labeling with the number of whites contained in the jar and the date. Angel cakes made with egg whites which had been frozen were judged to be as high in quality as those made with fresh egg whites, and frozen storage as long as six months did not influence the quality.

There is a distinct difference between yolks and whites in their ability to undergo freezing without detrimental changes. Yolks, in contrast to whites, become partially coagulated and gummy when frozen without additives. The problem can be overcome by adding either 10 percent salt or sugar prior to freezing the yolks. The additive needs to be selected to be compatible with the projected use of the yolk upon thawing so the flavor will be acceptable in the finished product. When frozen with salt or sugar at this level, yolks perform very satisfactorily, upon thawing, in sponge cakes (where they are beaten to a viscous foam) and in thickened products, such as custards.

The white of hard-cooked eggs is not satisfactory after freezing and thawing, for the white becomes distinctly rubbery. The yolk, however, undergoes freezing very well in a hard-cooked egg. Thus, the performance in the cooked state is just the reverse of that in the raw state. Apparently the problem in the whites develops as a result of the formation of large ice crystals in the white during frozen storage. These crystals create a porous texture and a watery character in the thawed hard-cooked egg white. Because of these adverse changes, casseroles or other dishes to be frozen should not include hard-cooked egg white in their formula.

Egg-thickened products are not satisfactory if they have been cooked prior to frozen storage.

As an example, baked custard curdles and exhibits syneresis following frozen storage. Similar results can be observed in cream puddings or egg-thickened fillings in cakes. A possible approach to solving these problems may be provided by using rice flour or rice starch as the thickening agent, but even this rather drastic approach does not completely solve the problem, and of course, flavor and color changes result if the egg is deleted. The more practical approach is to plan to use these types of products without freezing.

Egg white foams undergo freezing very satisfactorily in angel cakes and other types of cakes where they have been incorporated with other ingredients before baking. However, baked meringues toughen and shrink as a result of frozen storage. Because of these negative quality features, the freezing of baked meringues cannot be recommended. Similarly, seven-minute icings (also containing a cooked egg white foam) do not undergo freezing well. The thawed icing will be frothy and spongy, rather than the desired fluffy texture.

### Combinations of foods

Casseroles and some other recipes involve several different foods combined into one product. When considering their suitability for freezing, it is necessary to evaluate the ability of the various ingredients to undergo freezing and also to note whether treatment following frozen storage may be able to rectify problems. In casseroles, sauces may present an unattractive appearance when thickened with flour or cornstarch. Rice flour can be used to circumvent this problem reasonably well. Sauces which will be stirred and reheated are not of concern because the heating and stirring will reverse the textural problem created by freezing the starch-thickened mixture.

Some combinations of foods include mayonnaise. These recipes are not suitable for freezing because the emulsion in the mayonnaise will break as a result of freezing and, upon thawing, there will be a curdled appearance because of the separation of the oil.

Cooked meats and boiled noodles or rice often are included in casseroles. These undergo freezing very satisfactorily. Cheeses, onions, and many other ingredients added for flavoring also can be frozen in the casserole. These casseroles usually are of satisfactory quality when frozen either before or after baking. This makes it practical to prepare a double recipe at one time, using one when baked fresh and freezing the other one before it is baked. By placing a layer of aluminum foil in a casserole dish before placing the ingredients in the dish, the casserole can be frozen and then the aluminum-wrapped food can be removed from the dish for long-term storage. This releases the dish for use while the food is being stored.

Leftovers can be frozen for later use, too. Meat loaf is one meat item which is better if it is baked before being frozen. The baked meat loaf freezes very well.

Legume dishes freeze very well and can provide considerable convenience over preparing them from the basic ingredients each time they are prepared. Baked beans, for example, can be prepared in a large quantity; the desired amount for one meal can be eaten when freshly prepared, and the remainder can be packaged in meal-sized portions and frozen for later use. This approach can provide considerable saving in time and may make the difference between deciding to buy the prepared, canned product or the dried legumes and preparing them in the home. This latter decision provides good economy in meal planning when combining this choice with the use of the freezer for leftovers.

## Baked Products

Breads and rolls undergo freezing very successfully, particularly when frozen very promptly after baking and cooling. The richer formulas are less modified by slow rates of freezing than are the lean formulas, such as are used in French bread. However, the richer formulas, such as sweet rolls and Danish pastries, will undergo detectable deteriorative changes (particularly flavor changes) unless frozen storage is maintained at 0°F or lower. Seemingly, the higher fat content of these breads makes them more susceptible to oxidative changes when frozen storage temperatures are higher than the recommended level of 0°F. Sugar content may also be related to the observed staling under poor frozen storage conditions.

Cakes of all types can be frozen with excellent results following baking and cooling. All cakes can be wrapped and frozen when they have cooled to room temperature, and they should not stand at room temperature more than approximately 12 hours before freezing if optimum quality is to be retained. Freshly frozen pound cakes can be distinguished from unfrozen fresh pound cakes because of a detectable difference in firmness. However, pound cakes stored in the freezer for at least five weeks become more tender and, in fact, cannot be distinguished from freshly baked, unfrozen pound cakes. Other batter cakes store well at first when they are frozen and held at temperatures of 0°F or below, but their texture becomes harsh if held at warmer freezer temperatures. During the early part of storage at 0°F or below, batter cakes become progressively more moist and softer. This change is considered desirable for periods of storage as long as three or four months, but eventually the crumb will be perceived as being too moist and gummy. However, this long period of acceptable storage makes batter cakes particularly popular for freezer storage. Foam cakes undergo a very gradually increasing harshness and crumbliness during frozen storage regardless of the temperature of the freezer. This is in direct contrast to the batter cakes in which the difference in freezer temperature between 10°F and 0°F meant the difference between excellent storage characteristics and detectable loss of

*Figure 11.19. Casseroles made with cooked pasta freeze well. The food can be placed in a foil-lined casserole dish for freezing, wrapped tightly, and frozen in the dish. When frozen solid, the food can be stored in its aluminum foil wrapping without the dish. To prepare for serving, the frozen product is unwrapped and returned to the original casserole dish for heating and serving.*

quality in a relatively brief period (0°F being the temperature essential for long-term storage). However, chiffon cakes can be stored at least a month at 0°F very satisfactorily. Angel cakes also undergo textural changes over a period of time in frozen storage. At the very first, angel cakes develop an undesirable, tough crumb, but this gradually is transformed into a more moist, tender crumb which eventually is interpreted as bread-like. Three months of storage at 0°F appears to be an appropriate term of storage with retention of quality at an acceptable level for angel cakes.

Any of the batter or foam cakes can be frozen iced if a suitable icing is selected. As noted previously, seven-minute icings and egg-thickened fillings are not well suited for freezing, but cakes with butter icings are of excellent quality.

Fruit pies can be frozen prior to baking and held in frozen storage at a temperature of 0°F very satisfactorily. The unbaked crust is more resistant to physical damage in the freezer than is a baked crust, which actually will crumble or shatter rather easily if bumped in the freezer. Storage at a temperature of 0°F or lower is important as an aid in avoiding soaking of the filling into the bottom crust. Soaking of the crust is more of a concern in cherry pies and other juicy fillings than in less fluid fillings such as apple. Unbaked fruit pies can be held in frozen storage at 0°F for at least nine months with careful temperature control or for eight months if the freezer temperature fluctuates because of frequent opening of the door. Fruit pies, when frozen after being baked, are likely to have soggy bottom crusts.

*Figure 11.20 Butter icings freeze satisfactorily on cakes. These are the icings of choice if the iced product is to be frozen.*

*Figure 11.21. Croissants and other breads made from rich doughs can be frozen successfully after baking. The temperature for storage needs to be at least 0°F. or colder to prevent rapid deteriorative change due to the high fat content.*

*Figure 11.22. Chiffon cakes can be stored in the freezer for a month. It often is practical to freeze the cake without a topping and then add ice cream, a sauce, or whipped cream at the time of service.*

## Frozen Storage

Storage times for foods vary significantly from one type of food to another. Foods with a high fat content have a distinctly shorter freezer storage life than foods which do not contain fat because the fat will undergo slow oxidative changes during frozen storage. Pork provides a useful example of the limitations on freezer storage due to fat content. The recommended maximum lengths of time for frozen storage are presented in Table 11.3.

One of the problems influencing quality in frozen foods is the negative influence fluctuations in temperature have on frozen items. Partial thawing and refreezing causes considerable cellular disruption in fruits, vegetables, and meats. Another potential problem during frozen storage is the development of freezer burn. This actually is a matter of dehydration or desiccation rather than burning. When food is not tightly wrapped to exclude the air of the freezer from coming in contact with the surface of the food, there will be a dehydration of the cells at the surface due to the extremely dry environment of the freezer. Freezer burn can be avoided completely by wrapping or packaging foods so that they are virtually air tight.

Table 11.3. Suggested maximum freezer storage for selected foods

| Food | Approximate holding period at 0°F (months) | Food | Approximate holding period at 0°F (months) |
|---|---|---|---|
| **Fruits and Vegetables** | | **Meat (continued)** | |
| Fruits: | | Cooked meat: | |
| Cherries | 12 | Meat dinners | 3 |
| Peaches | 12 | Meat pie | 3 |
| Raspberries | 12 | Swiss steak | 3 |
| Strawberries | 12 | **Poultry** | |
| Fruit juice conc.: | | Chicken: | |
| Apple | 12 | Cut-up | 9 |
| Grape | 12 | Livers | 3 |
| Orange | 12 | Whole | 12 |
| Vegetables: | | Duck, whole | 6 |
| Asparagus | 8 | Goose, whole | 6 |
| Beans | 8 | Turkey: | |
| Cauliflower | 8 | Cut-up | 6 |
| Corn | 8 | Whole | 12 |
| Peas | 8 | Cooked poultry: | |
| Spinach | 8 | Dinners (chicken or | |
| **Baked Goods** | | turkey, gravy) | 6 |
| Bread, yeast rolls: | | Poultry pies | 12 |
| White bread | 3 | Fried chicken | 4 |
| Cinnamon rolls | 2 | Fried chick. dinner | 4 |
| Plain rolls | 3 | **Fish and Shellfish** | |
| Cakes: | | Fish: | |
| Angel | 3 | Fillets: | |
| Chiffon | 2 | Cod, flounder, had- | |
| Chocolate layer | 4 | dock, halibut, | |
| Fruit | 12 | pollack | 6 |
| Pound | 6 | Mullet, ocean | |
| Yellow | 6 | perch, sea bass | 2 |
| | | Sea trout, | |
| Danish pastry | 3 | striped bass | 3 |
| Doughnuts: | | Pacific perch | 2 |
| Cake type | 3 | Salmon steaks | 2 |
| Yeast raised | 3 | Sea trout, dressed | 3 |
| Pies (unbaked): | | Striped bass, dr. | 3 |
| Apple | 8 | Whiting, drawn | 4 |
| Boysenberry | 8 | Shellfish: | |
| Cherry | 8 | Clams, shucked | 3 |
| Peach | 8 | Crabmeat: | |
| **Meat** | | Dungeness | 3 |
| Beef: | | King | 10 |
| Hamburger, chipped | | Oysters, shucked | 4 |
| (thin) steaks | 3 | Shrimp | 12 |
| Roasts | 12 | Cooked fish, shellfish: | |
| Steaks | 12 | Fish, cheese sauce | 3 |
| Lamb: | | Fish, lemon butter | |
| Patties (ground meat) | 3 | sauce | 3 |
| Roasts | 12 | Fried fish dinner | 3 |
| Pork, cured | 2 | Fried fish sticks, | |
| Pork, fresh: | | scallops, shrimp | 3 |
| Chops | 4 | Shrimp creole | 3 |
| Roasts | 8 | Tuna pie | 3 |
| Sausage | 2 | **Frozen Desserts** | |
| Veal: | | Ice cream | 1 |
| Cutlets, chops | 4 | Sherbet | 1 |

## SUMMARY

Freezing is a convenient method of food preservation which can be done easily in the home with little special equipment. For long-term storage, a freezer is necessary so that the temperature can be maintained consistently at 0°F or below. However, the freezer unit of a refrigerator can be used for short-term frozen storage very satisfactorily. Packaging of frozen foods is done using plastic containers, glass jars, or air-tight, moisture-resistant wrapping such as heavy aluminum foil or plastic bags. Not only is the equipment available to most people, but the short time required to prepare items for the freezer also fits the resources available to many. Another advantage of freezing is that the quality of many items when thawed after frozen storage approximates the quality of the freshly cooked food.

Fruits are frozen using either a sugar syrup, a dry sugar pack, or no added sugar. The texture of fruits will be modified by freezing, making it appropriate to freeze fruits which will be used in sauces or cooked products rather than freezing them for use as the uncooked fruit. The texture of melons is affected much more negatively than is true for most other fruits. Browning of fruits can be prevented or minimized by preventing oxidation of pigments by adding an antioxidant such as ascorbic acid and packaging fruits to insure that they are held beneath the surface of the syrup in which they are packed.

The enzymes in vegetables will cause deteriorative changes in pigments and flavors if the vegetables are not blanched prior to freezing. The recommended practice is to blanch and immediately cool and drain vegetables for prompt packaging and freezing. Vegetables are packed without liquid or salt.

The decision to freeze quantities of meats, fish, or poultry is based largely on the availability and economy provided by freezing them. In any event, it is important to calculate how much might reasonably be expected to be consumed in a reasonable length of time so that the food will be consumed before the recommended maximum storage period has elapsed. Careful attention to thawing of these items as well as quick preparation for freezing must be given to these foods since they represent potential avenues for food-borne infections if they are allowed to remain at room temperature for any extended period of time.

Recipes featuring mixtures of foods need to be examined to be certain that the ingredients included are well suited to freezing. Starch-thickened foods can be frozen satisfactorily if they will be reheated and stirred following freezing. Egg yolks require added sugar or salt if they are to be frozen raw, but they are of excellent quality when frozen without additives after they are hard-cooked. In contrast, egg whites freeze well when uncooked, but hard-cooked egg whites are watery and porous. Baked products generally undergo freezing with excellent results. Cakes are baked frozen, but fruit pies are of better quality when baked following frozen storage.

An important routine is to note the operation of a freezer frequently to be certain that the necessary temperature of 0°F is being maintained. Attention needs to be given to being certain that the freezer is defrosted frequently enough to insure maintenance of this low temperature. Additional good management practices for freezer operation include careful, clear labeling of all foods placed in the freezer and a diagram of the foods in the freezer so that good rotation and quick location of the desired product can be accomplished.

## BIBLIOGRAPHY

Animal Husbandry Research Div., et al. *Freezing Meat and Fish in the Home.* Home and Garden Bulletin No. 93. U.S. Dept. Agriculture. Washington, D.C. 1970.

Consumer and Food Economics Institute. *Home Freezing of Poultry and Poultry Main Dishes.* Agriculture Information Bulletin No. 371. U.S. Dept. Agriculture. Washington, D.C. 1975.

Consumer and Food Economics Research Div. *Home Care of Purchased Frozen Foods.* Home and Garden Bulletin No. 69. U.S. Dept. Agriculture. Washington, D.C. 1971.

Consumer and Food Economics Research Div. *Home Freezing of Poultry.* Home and Garden Bulletin No. 70. U.S. Dept. Agriculture. Washington, D.C. 1970.

Desrosier, N.W. *Fundamentals of Food Technology.* Avi. Westport, Conn. 1977.

Desrosier, N.W. and J.N. Desrosier. *Technology of Food Preservation.* Avi. Westport, Conn. 4th ed. 1977.

Desrosier, N.W. and D. K. Tressler. *Fundamentals of Food Freezing.* Avi. Westport, Conn. 1977.

Dawson, L.E. Stability of frozen poultry meat and eggs. In *Quality and Stability of Frozen Foods.* Ed. W.B. Van Arsdel, et al. Wiley Interscience. New York. 1969.

Fields, M.L. *Laboratory Manual in Food Preservation.* Avi. Westport, Conn. 1977.

Groppe, C.C., et al. *Canning and Freezing Fish at Home.* Cooperative Extension, University of California. Berkeley. 2nd ed. 1974.

Guadagni, D.G. Quality and stability in frozen fruits and juices. In *Quality and Stability of Frozen Foods.* Ed. by W.B. Van Arsdel, et al. Wiley Interscience. New York. P. 85. 1969.

Jansen, E.F. Quality-related chemical and physical changes in frozen foods. In *Quality and Stability of Frozen Foods.* Ed. W.B. Van Arsdel, et al. Wiley Interscience. New York. P. 19. 1969.

Joslyn, M.A. *Bottling, Canning, Freezing Fruit Juices and Tomato Juice.* Family and Consumer Sciences Leaflet 4002. Cooperative Extension, University of California. Berkeley. 1974.

Michener, H.D. and R. P. Elliott. Microbiological conditions affecting frozen food quality. In *Quality and Stability of Frozen Foods.* Ed. W.B. Van Arsdel, et al. Wiley Interscience. New York. P. 43. 1969.

Olson, R.L. and W.C. Dietrich. Quality and stability of frozen vegetables. In *Quality and Stability of Frozen Foods.* Ed. W.B. Van Arsdel et al. Wiley Interscience. New York. P. 117. 1969.

Pence, J.W. Quality and stability of frozen bakery products. In *Quality and Stability of Frozen Foods.* Ed. W.B. Van Arsdel, et al. Wiley Interscience. New York. P. 169. 1969.

Tressler, D.K. *Freezing of Precooked and Prepared Foods.* Avi. Westport, Conn. 1968.

Van Arsdel, W.B. *Factors Affecting Quality in Frozen Foods.* Wiley Interscience. New York. 1969. zen Foods. Avi. Westport, Conn. 1968.

Van Arsdel, W.B. *Quality and Stability of Frozen Foods.* Wiley Interscience. New York. 1969.

*Figure 11.23. Tight wrapping of meats is important in preventing freezer burn. A piece of aluminum foil large enough to wrap the meat with at least two inches of overlap is needed (upper). The edges of foil are brought together and folded down to the surface of the meat with two folds to seal out the air (middle). The wrap is completed by folding the ends tightly two times (lower). After labeling, the meat is ready for freezer storage.*

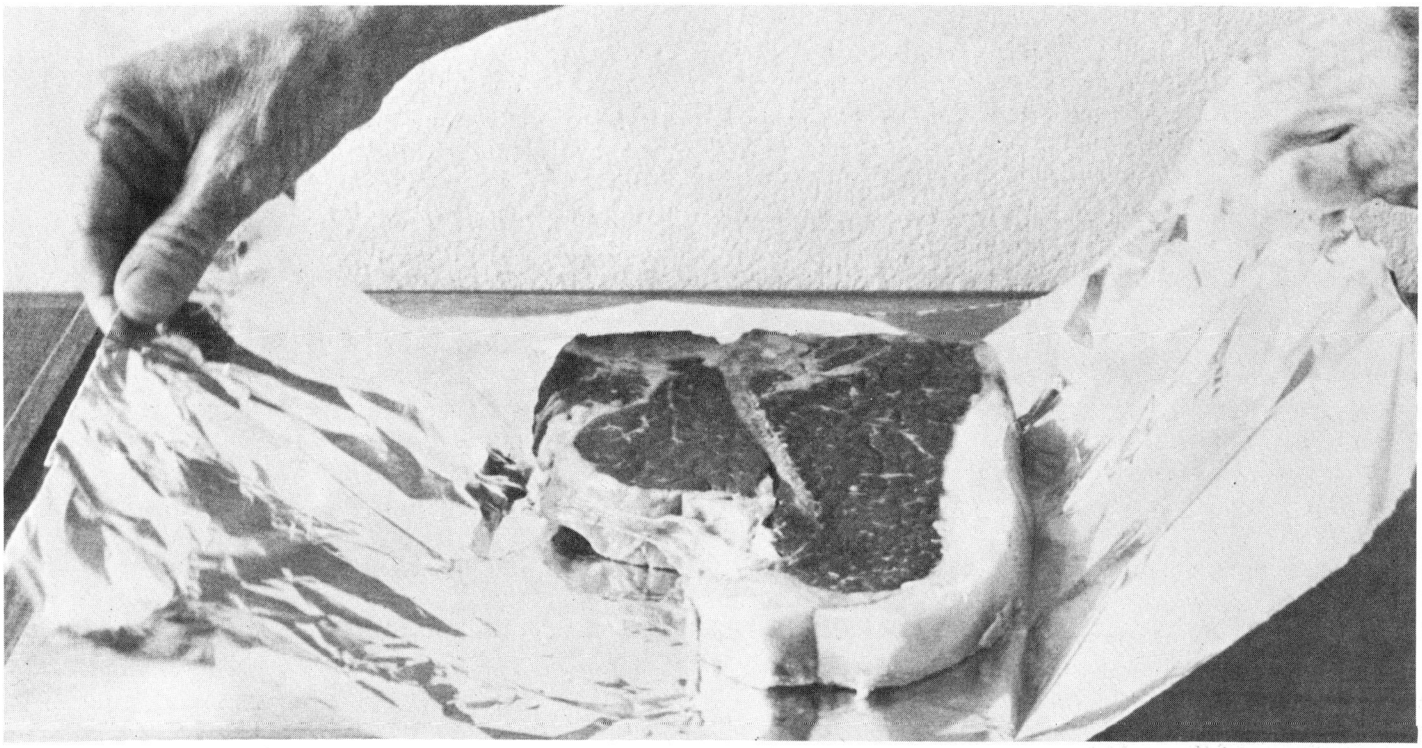

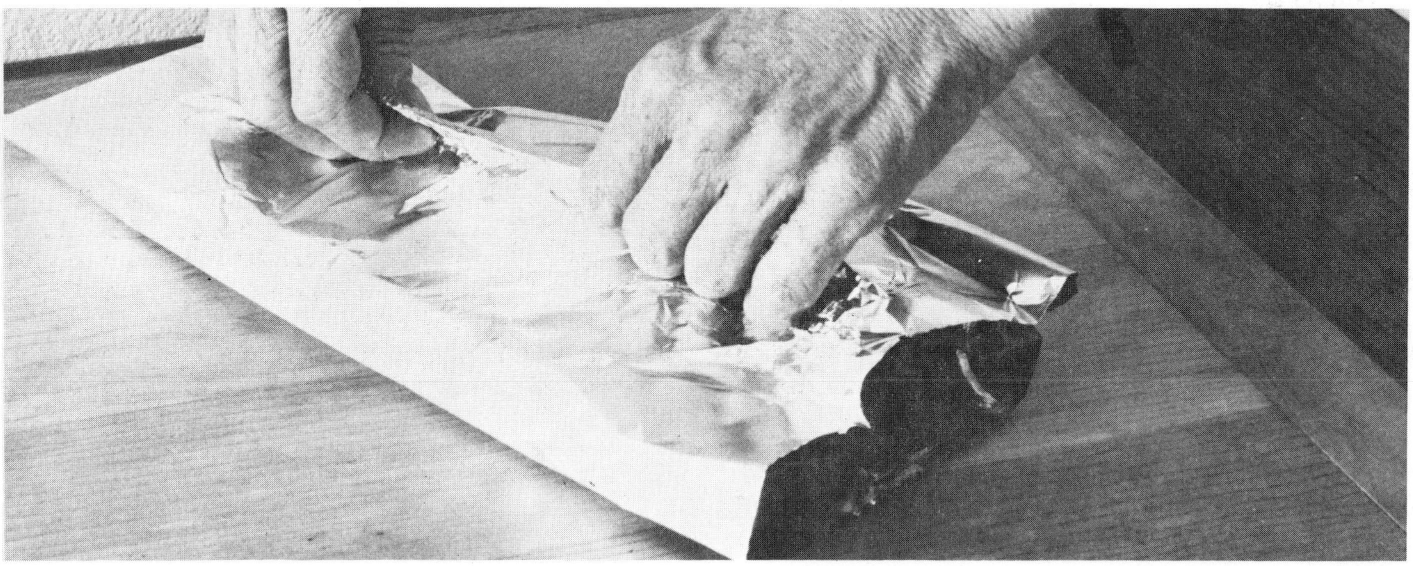

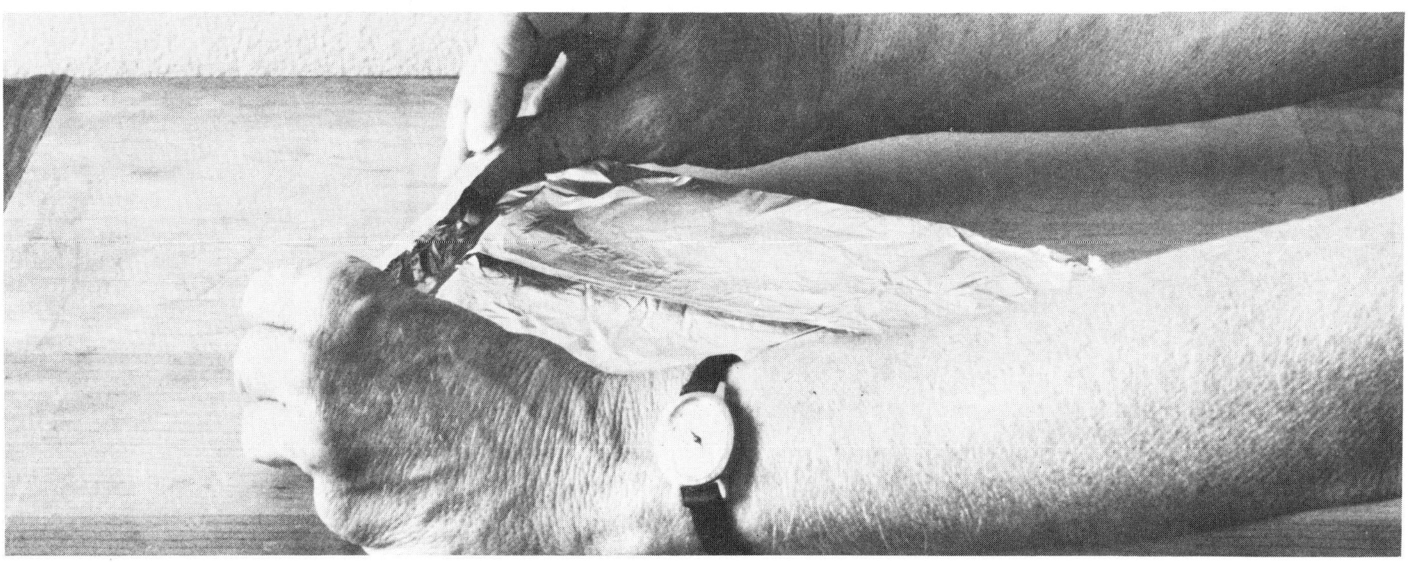

# Chapter 12. Recipes for Freezing

The possibilities available for freezing foods as a means of preserving them seemingly are endless. This chapter is planned to provide an introduction into the techniques involved in freezing basic products and to reveal some of the potential for increasing the range of foods included in the diet and for refining techniques of time and money management through use of the freezer. From the principles illustrated here, each person will be able to begin to evolve an individualized approach to freezing as a versatile and convenient means of preserving foods.

## FRUITS

### Basic Freezing Techniques

As described in Chapter 11, fruits can be packed in a sugar syrup, a dry sugar pack, or without sugar. The most common method of packing fruits is packing with the sugar syrup. The procedures for freezing by these three methods are outlined below.

The sugar syrups used for packing many fruits to be frozen need to be made prior to preparing the fruits for freezing so that the syrup will be very cold when added to the fruit. This measure will speed the freezing of the fruit in the freezer because it will not be necessary to remove as much heat from the syrup as would be required if the syrup were warmer at the beginning of the freezing period. Very fast freezing is an aid in avoiding very large crystals which cause considerable cellular disruption when they form in freezing foods.

The concentration of the syrup used in the freezing of fruits is a matter of individual choice. Of course, syrups with a higher concentration of sugar will be sweeter in flavor; they also will be more viscous than syrups containing a smaller percentage of sugar. The approximate amounts of sugar and water for making syrups of varying concentrations are presented in Table 12.1. Very sweet fruits can be prepared satisfactorily using a 30 percent sugar syrup, while very tart foods, such as cranberries, may require up to 60 percent syrups.

Table 12.1. Formulas for sugar syrups

| Syrup | Sugar (cups) | Water (cups) | Kind of syrup |
|---|---|---|---|
| 30% | 2 | 4 | Medium |
| 40% | 3 | 4 | Medium heavy |
| 50% | 4 | 4 | Heavy |
| 60% | 6 | 4 | Extra heavy |

**Strawberries (sugar syrup pack)**

3 quarts strawberries
2 cups sugar
2 cups water

1. In advance of the actual preparation of the berries for freezing, heat the sugar and water together in a saucepan and boil the syrup until the sugar is dissolved. Cool thoroughly before beginning the preparation of the berries.
2. Wash strawberries carefully; hull the berries and cut away any soft or spoiled areas; slice into a bowl or leave whole, depending on personal preference.
3. After the fruit has drained well, pack the berries into plastic containers or freezer jars, being careful to pack the fruit efficiently, but not to bruise it. Leave an inch of headspace.
4. Pour the prepared syrup (50 percent) over the fruit in each container so that the fruit is covered by the syrup and half an inch of headspace is left to allow for expansion during freezing.

5. Crumple some wax paper or plastic wrap and place on the fruit.
6. Wipe the rim of the container clean before placing the closure tightly on the container. The crumpled wrap and the lid should serve to hold the berries under the surface of the syrup to aid in preventing oxidation of the fruit and improve the quality of the thawed product following frozen storage.
7. Label the containers clearly with the name of the product and the date of freezing.
8. Immediately place the berries in the freezer, spreading them apart so that there is good circulation between them to facilitate freezing. It also will speed freezing if the freezer is turned to its coldest setting until the berries are frozen.

**Strawberries (dry sugar pack)**

3 quarts strawberries
2 cups sugar

1. Wash strawberries carefully; hull the berries and cut away any soft or spoiled areas as was done for the syrup pack.
2. Slice the berries or leave them whole, as preferred, and place the well drained berries in a large bowl.
3. Stir the sugar very gently into the berries until all of the berries are coated with sugar.
4. Pack the sugar-coated fruit into the containers, leaving half an inch of headspace. Pour the juice remaining in the bowl into the containers. There should be about enough juice to cover the fruit.
5. Proceed with the closure, labeling, and freezing of the fruit in the same way as described above.

**Strawberries (no sugar)**

3 quarts strawberries

1. Wash strawberries carefully; hull the berries and cut away any soft or spoiled areas as was done for the preceding methods of packing.
2. Slice the berries or leave them whole, as preferred, and place the berries (well drained) directly into the freezing containers.
3. Without the addition of any sugar or syrup, simply close the containers tightly, label, and freeze the fruit as described above.

**Peaches (using ascorbic acid)**

7 pounds peaches
1 quart chilled sugar syrup (30%)
¾ teaspoon ascorbic acid

1. Wash peaches well before dipping in boiling water just long enough to loosen the skins.
2. Plunge peaches into ice water just long enough to cool them.
3. Peel the peaches quickly; slice into a bowl of water to which a tablespoon of lemon juice has been added.
4. When all the peaches have been prepared, add the ascorbic acid to the chilled syrup and stir to dissolve it.
5. Drain the peaches and gently pack them into the containers.
6. Pour the syrup over the fruit to just cover, being sure that half an inch of headspace remains.
7. Crumple plastic wrap or wax paper, position over the fruit to hold the slices under the sugar syrup, and position the cover securely.
8. Label and place in the freezer, leaving room between each container. Be sure the freezer has been turned to its coldest setting for freezing and that it subsequently is turned back to normal temperature to maintain a temperature of 0°F or below.

**Other fruits**

The above directions for strawberries and peaches outline the general procedures to be followed when preparing fruits for frozen storage. The decision points are: (1) whether to pack in syrup or to dry sugar pack, (2) how sweet to make the product, and (3) whether ascorbic acid (or other additive) is needed to prevent discoloration of the fruit. These decisions are primarily a matter of personal preference. In addition to making these decisions, the amount to be frozen should be kept in mind. Freezers in the home can freeze new foods more quickly if a small amount is placed in the freezer at one time. Best results will be obtained if only about 6 or 7 pints are processed at a time. Suggestions for preparing some of the fruits often frozen at home are presented in Table 12.2.

## Fruit Specialties

**Frozen bananas**

Ripe bananas

1. Peel ripe bananas.
2. Immediately wrap in plastic wrap or heavy aluminum foil, being careful to wrap so that air is excluded as much as possible from the banana.
3. Label and place in freezer. Use within a month.

*Note:* To serve, slice very thinly into sherbet glasses and top with a dollop of whipped cream or dairy sour cream. Sprinkle with ground cinnamon or nutmeg. Work quickly so that bananas will still contain ice crystals when served.

**Banana Frosties**

3 large ripe bananas
8 oz. semi-sweet chocolate (or milk chocolate)
Few curls of paraffin
1 c. chopped nuts or flaked coconut

1. Peel bananas, cut in 1 inch slices, and insert a toothpick in each.

Table 12.2. Guidelines for freezing fruits

| Fruit | Preparation | Type of pack |
|---|---|---|
| Apricots | Wash, cut in halves, pit. | 40% syrup, ¾ tsp. asc. acid; ½ c. sugar/qt. |
| Avocados | Use soft fruit. Peel, pit, mash. Add 2 tsp. lemon juice/fruit | ⅛ tsp. salt/qt. no sugar |
| Cherries, pie | Wash. Remove stems and pits. | 60% syrup |
| Cherries, sweet | Wash. Remove stems and pits leave whole. | 40% syrup; or ¾c. sugar/qt. |
| Cranberries | Wash, sort, stem. | 50% syrup; or no sugar |
| Nectarines | Wash, peel, pit, slice into 1 qt. water and 1 tbsp. lemon juice | 30% syrup, ½ tsp. asc. acid |
| Pears | Wash, peel, quarter, core, slice. | 40% syrup, ¾ tsp. asc. acid |
| Plums | Wash. Halve and pit or leave whole. | 50% syrup; or pack dry |
| Raspberries | Wash and sort carefully. | 40% syrup; or ¾ c. sugar/qt. |
| Rhubarb | Wash, trim, cut in 1-in. pieces | 50% syrup |

2. Arrange on foil on a small cookie sheet, and place on floor of freezer section until firm. Do not cover or wrap the banana sections.
3. Melt chocolate and paraffin over boiling water.
4. Dip frozen banana sections in melted chocolate; roll in nuts or coconut.
5. Serve immediately or arrange carefully in plastic container and cover for freezer storage.

*Note:* Butterscotch chips and 1 tbsp. butter may be substituted. Peanuts are a good choice with this.

**Peaches in Orange Juice**

1 can (6 oz.) frozen orange juice concentrate
3 cans water
½ cup sugar (optional)
10 medium peaches (about 3 pounds)

1. Mix orange juice with water. Add sugar, if used.
2. Pour orange juice into freezer containers until they are about one third full.
3. Wash, peel, pit, and slice peaches directly into the juice in the containers. Leave ½ inch headspace.
4. Add orange juice to cover peach slices, leaving ½ inch headspace.
5. Crumple wax paper or plastic wrap to hold fruit under the juice, and close the cover securely.
6. Label and freeze.

*Note:* Partially thaw in the container, but serve while fruit still has ice crystals.

**Mixed Fruit Cooler**

3 cups sugar
3 cups water
1 package (10 oz.) frozen strawberries
1 package (10 oz.) frozen raspberries (optional)
Juice of 3 oranges
Juice of 3 lemons
3 bananas cubed
3½ cups (28-oz. can) pineapple chunks with juice
2 quarts ginger ale (or other sparkling beverage)

1. Heat sugar and water to a boil, stirring to dissolve the sugar.
2. Remove syrup from heat, and add frozen berries. Stir berries occasionally to aid in defrosting.
3. When syrup is cool and berries have thawed, add fruits and juices and stir gently to mix well.
4. Place mixture in plastic container large enough to allow use of an ice cream scoop for removing portions following frozen storage. Cover container tightly and freeze.

*Note:* Serve by placing a scoop of the frozen mixture in a sherbet glass and adding ginger ale to just cover. Let this thaw slightly before serving as an appetizer cup or dessert.

Yield: 20 servings.

**Cranberry Ribbon Loaf**

1 lb. can whole cranberry sauce
2 tablespoons lemon juice
½ pint whipping cream
¼ cup powdered sugar
⅔ cup chopped nuts
1 can (8 oz.) crushed pineapple, drained

1. Combine cranberry sauce with lemon juice and pour into loaf pan.
2. Whip the cream until stiff.
3. Fold the sugar, nuts, and pineapple into the whipped cream.
4. Carefully transfer the whipped cream mixture to make a layer on the cranberry layer. Spread smoothly.
5. Cover the loaf pan tightly and place in freezer.
6. Freeze until solid.

*Note:* Serve in slices for salad or dessert. If frozen storage will be more than a few days, line the loaf pan with foil before adding the ingredients. This allows the frozen loaf to be removed easily from the pan and wrapped for storage without the pan.

**Chocolate Covered Fruit**

Seedless grapes
Bing cherries (preferably with stems)
Pineapple (cut in bite-sized cubes)
Strawberries
Semi-sweet or milk chocolate

1. Wash fruit well and drain thoroughly.
2. Over hot water, melt the chocolate until almost melted.
3. Remove chocolate from heat and stir until completely melted.
4. Dip fruit, one piece at a time, into the chocolate. Remove after rolling with a fork or thin skewer.
5. Let excess chocolate drip into pan of chocolate, and then place on aluminum foil to cool.
6. Chill until chocolate is firm.
7. Place in plastic freezer container, cover tightly, and freeze.

*Note:* Serve while still frozen. Any fruit may be prepared in this way.

**Snow Strawberries**

Whole, fresh strawberries
White chocolate

1. Wash berries carefully and drain well.
2. Over hot water, melt chocolate until soft. Stir to finish blending the melting chocolate.
3. Holding the cap carefully, dip each strawberry into the melted white chocolate so that about half of the berry is coated with chocolate.
4. Let chocolate-covered berries sit on foil until the chocolate hardens.
5. Transfer to plastic freezing container, cover tightly, and freeze.

*Note:* Serve berries while still frozen. This same procedure can be used for coating Bing cherries or seedless grapes.

**Frozen Fruit Salad**

1 cup whipping cream
¼ cup mayonnaise
2 bananas
1 cup fruit cocktail, drained
½ cup salad marshmallows
2 tablespoons sugar

1. Whip the cream until stiff.
2. Slice the bananas into the whipped cream and add the other ingredients.
3. Stir gently to thoroughly mix all ingredients.
4. Line loaf pan with aluminum foil, and then transfer the salad to the loaf pan.
5. Cover pan tightly with aluminum foil, label, and freeze.
6. After salad has frozen hard, remove from the pan and rewrap. Return to freezer for frozen storage.

*Note:* Slice and serve frozen. This is a convenient salad for holiday meals because it can be made ahead. It also is convenient to serve unexpected guests.

## VEGETABLES

### Basic Freezing Techniques

Freezing of vegetables is a relatively simple process requiring little time. However, it is important to be certain that the blanching and subsequent cooling are done with careful attention. This step is the key to successful freezing of vegetables.

**Green beans**

6 pounds green beans

1. Begin heating a gallon of water to a boil in a large kettle.
2. Wash beans carefully. Cut off both ends; leave whole, cut into 1-inch segments, or cut in julienne style.
3. Prepare large bowl or pan of ice water large enough to immerse the basket which will be used for blanching.
4. Place approximately a quart of the prepared beans in a metal basket, lower the basket into the rapidly boiling water in the kettle. Place the cover on the kettle to help trap the heat, and blanch the beans for 3 minutes.
5. Just as soon as the blanching period is over, remove the basket of beans and plunge it into the ice water, leaving in the ice water for 3 minutes.
6. Drain the chilled beans, and pack efficiently into freezer containers, leaving half an inch of headspace.
7. Crumpled plastic wrap may be placed on top of the beans to fill the headspace and reduce formation of ice crystals on the surface of the beans.
8. Cover the containers tightly, label, and freeze.

**Other vegetables**

The procedures for preparing vegetables for freezing are very similar for all vegetables. There are some variations in the preparation and in the blanching period. As was true with fruits, the amount of vegetable to be frozen at one time should be limited ideally to about 6 or 7 pints so that the freezer will be able to maintain the low temperature needed for very rapid freezing. This is an aid in assuring small crystals of ice and minimum cell destruction during freezing.

The specific suggestions for freezing selected vegetables are presented in Table 12.3.

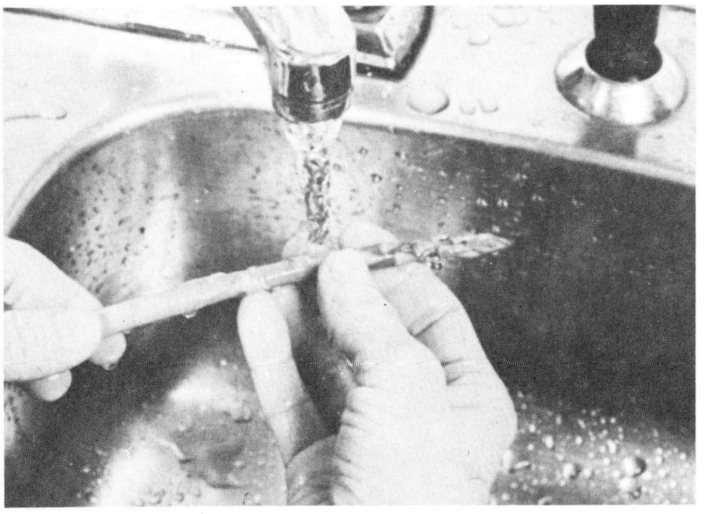

*Figure 12.1 Asparagus, like all other vegetables, must be washed very thoroughly under running water.*

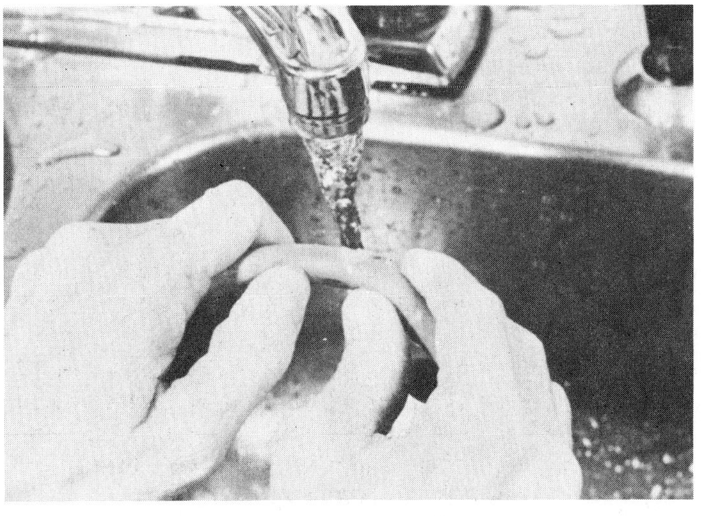

*Figure 12.2. It is quick and practical to test the asparagus to see where it will bend and just begin to break. This is the extent of the vegetable which will be tender.*

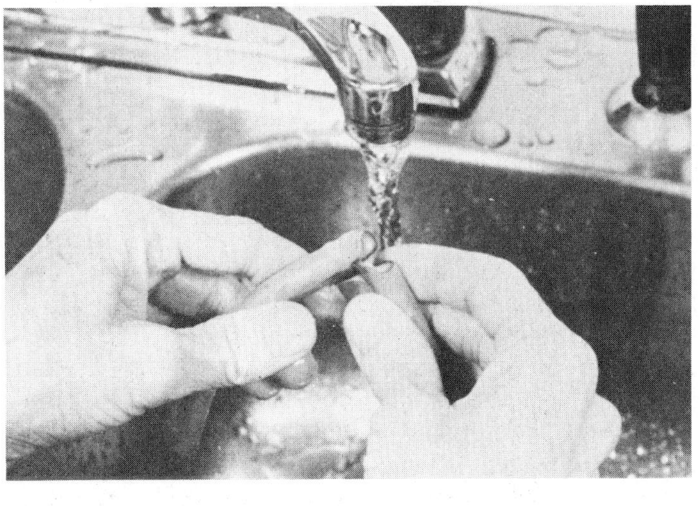

*Figure 12.3. Snap off the asparagus stalk at the bend, and discard the woody portion.*

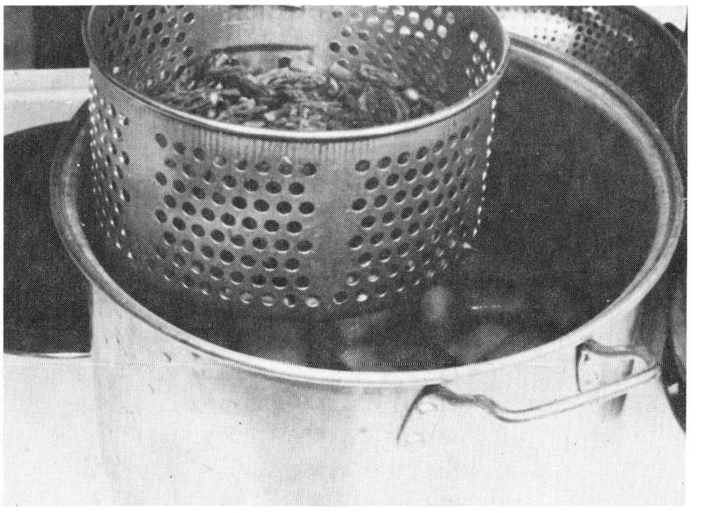

*Figure 12.4. Asparagus is blanched and then immediately cooled in ice water in preparation for freezing.*

Table 12.3. Guidelines for freezing selected vegetables

| Vegetable | Preparation | Blanching time (min.) |
|---|---|---|
| Asparagus | Wash thoroughly- break off tough stalk; cut in spears or 1" pieces | Small - 2<br>Medium - 3<br>Large - 4 |
| Beans, lima | Shell and wash | Small - 2<br>Large - 4 |
| Broccoli | Wash and trim well, removing leaves and insects. Split lengthwise | 3 |
| Brussels sprouts | Trim, remove outer leaves. Wash | Small - 3<br>Large - 5 |
| Cauliflower | Wash. Cut into flowerets. Trim | 3 |
| Corn | Husk, wash. Cut kernels from cob | 4 |
| Greens | Wash very well. Remove stems | 3 |
| Peas | Shell. Sort. Wash | 2 |
| Squash, winter | Wash, remove seeds, and cook in small pieces till tender. Puree. Chill | - |
| Pumpkin | Wash. Remove seeds. Cook in small pieces till tender. Puree. Chill | - |

## Vegetable Specialties

### Copper Pennies

2 pounds carrots, peeled, sliced ¼ inch thick
1 green bell pepper, coarsely chopped

1 can (10 oz.) tomato soup, undiluted
¾ cup sugar
½ cup vinegar
½ cup oil
1 teaspoon salt
1 teaspoon dry mustard
1 teaspoon Worcestershire sauce

1. Boil carrots in salted water to just barely cover until just tender. Drain and cool.
2. Heat the soup, sugar, vinegar, oil, and seasonings to a boil and boil for about 2 minutes.
3. Combine carrots and green pepper; pack in freezer containers, leaving ¾ inch headspace.
4. Pour cooled tomato soup marinade over the carrots to cover the carrots. Add a crumpled piece of plastic wrap and cover tightly.
5. Label and freeze.

*Note:* Copper Pennies may be thawed and served either hot as a vegetable or cold as a garnish. The extra marinade is a flavorful salad dressing. Small kabobs of carrot, green pepper, and cocktail onions arranged on a toothpick and served with the marinade as a dip make healthful hors d'oeuvres.

**Frozen Tomato Cocktail**

4 tomatoes
½ medium onion
1 green bell pepper (seeded)
1 carrot
1 teaspoon salt
1 tablespoon sugar

1. Wash the tomatoes, onion, pepper, and carrots. Peel and core the tomatoes, peel the onion, remove seeds and stem of the pepper, and pare carrot.
2. Cut the carrot into large pieces before adding it to the other ingredients which are placed in the blender.
3. Add the salt and sugar and whirl in the blender until all ingredients are pureed.
4. Heat mixture to boiling.
5. Immediately cool puree.
6. When cool, pour into freezer containers, label, and freeze.

*Note:* Thawed cocktail is served cold over ice cubes or diluted with ice water to the desired consistency. The undiluted puree is a flavorful base for spaghetti sauce or other dishes requiring a tomato base.

**Green Beans Supreme**

4 pounds green beans
2 packages mushroom gravy mix
8 cups water

1. Stir 1 cup water into mushroom gravy mix to make a slurry. Gradually stir in the remaining water.
2. While the gravy is being brought to a boil, wash the beans; remove both ends, and cut into diagonal slices about 1½ inches long.
3. Add the prepared beans to the boiling gravy and boil for 3 minutes.
4. Place ½ cup of the gravy in the freezer containers.
5. Use a slotted spoon to transfer beans to the freezer containers. Leave ½ inch headspace.
6. Add gravy to cover the beans. Place crumpled plastic wrap in the top of the container and close tightly.
7. Label and freeze.

*Note:* To serve, heat slowly to boiling, being careful to avoid burning while all of the sauce is melting.

**Corn Miniatures**

Ears of corn

1. Husk corn carefully, removing the silks.
2. Trim the ends and wash the ears well. Use a heavy knife to cut into pieces of the ear about 2 inches long.
3. Blanch the pieces for 7 minutes in boiling water.

*Figure 12.5. Green Beans Supreme are blanched in the mushroom gravy and the gravy subsequently is used for packaging. This saves vitamins.*

*Figure 12.6. Blanching in the gravy is done for the same length of time as would be used if blanching in boiling water.*

4. Plunge the basket of corn into ice water for 7 minutes to cool.
5. Drain ears well. Wrap several miniatures together in heavy aluminum foil or package in plastic pouches. Seal tightly, label, and freeze.

*Note:* Place foil-wrapped corn in 400°F oven for 45 minutes to thaw and cook prior to serving.

**Butter-Crumb Corn Patties**

1 can (1 lb.) whole kernel corn, drained
3 eggs, well beaten
1 cup grated cheddar cheese
2 tablespoons finely chopped onion
1 teaspoon salt
Dash pepper
1¼ cups finely crushed dry bread crumbs
Butter or margarine (for browning)

1. Combine all ingredients except ½ cup of the bread crumbs.
2. Shape into patties, using a ¼-cup measuring cup as a mold.
3. Roll each patty in the reserved bread crumbs, pressing the crumbs in to help them adhere to the patties.
4. Place individual patties on cookie sheet and place, uncovered, in the freezer until frozen.
5. Remove from the freezer; wrap in heavy foil or in a plastic container, using paper to separate individual patties. Label and return to freezer.

*Note:* To serve, fry while still frozen in butter or in margarine. The patties should be a pleasing, golden brown on the exterior and heated to serving temperature on the interior.

Yield: About a dozen patties.

**Frozen Onion Rings**

Spanish onions
Milk
Pancake mix (dry)
Cornmeal
Oil for deep fat frying

1. Wash onions. Slice into thick slices, separate rings, and slip off the peels.
2. Blanch a few rings at a time for 10 to 15 seconds in boiling water.
3. Plunge into ice water for equal length of time.
4. Drain well before shaking rings, a few at a time, in a bag of flour to coat the rings.
5. Dip the flour-coated rings in milk.
6. Shake a few rings at a time in a mixture of equal amounts of pancake mix and cornmeal.
7. Arrange the coated rings on cookie sheets. If necessary, wax paper can be placed on top of the layer of prepared onion rings and an additional layer placed on top.
8. Freeze the onions before transferring to plastic containers. Separate layers in the containers, using plastic wrap.
9. Return to freezer after labeling containers.

*Note:* Fry the rings in oil preheated to 375°F until they are golden brown. The rings, since they are frozen when placed in the oil, hold their shape and their breading well. Fry only the amount needed for the occasion, returning the remainder of the rings quickly to the freezer.

**Perfect Twice-Baked Potatoes**

Medium-sized Russet potatoes
Milk or half-and-half
Baking powder (¼ teaspoon/each 2 potatoes)
Seasonings to taste (butter, minced onion, bacon bits)

1. Scrub potatoes very well.
2. Bake at 400°F for an hour, until soft.
3. Cut a slice from the top of each potato large enough for the potato to be scooped easily from the shell.
4. Scoop the potato from the shell and place in a bowl for mashing, using a spoon to scoop out the potato. Care must be taken to be sure to avoid tearing the skin.
5. Place the shells on a cookie sheet and return to oven to crisp shells.
6. Put potatoes through a ricer or mash with a potato masher until smooth.
7. Add hot milk, baking powder, and desired seasonings, and whip until light and fluffy.
8. Remove shells from oven. Refill them with the mashed potato mixture, piling the filling attractively and sprinkling with paprika or grated cheese.
9. Let potatoes cool on cookie sheet. Then transfer the potatoes on the sheet (uncovered) to the freezer and freeze until solid before removing to wrap potatoes in aluminum foil, labeling, and returning to freezer.

*Note:* To serve these potatoes, unwrap them, place on a cookie sheet, and bake in a 350°F or 400°F oven until heated through thoroughly.

**Mashed Potato en Casserole**

6 medium Russet potatoes
2 tablespoons minced instant onion
1 cup sour cream
1 cup creamed cottage cheese
2 teaspoons baking powder
Salt and white pepper to taste

1. Bake potatoes at 400°F until soft. Scoop out the potato carefully, saving the skins for the Tater Peel Crispies recipe.

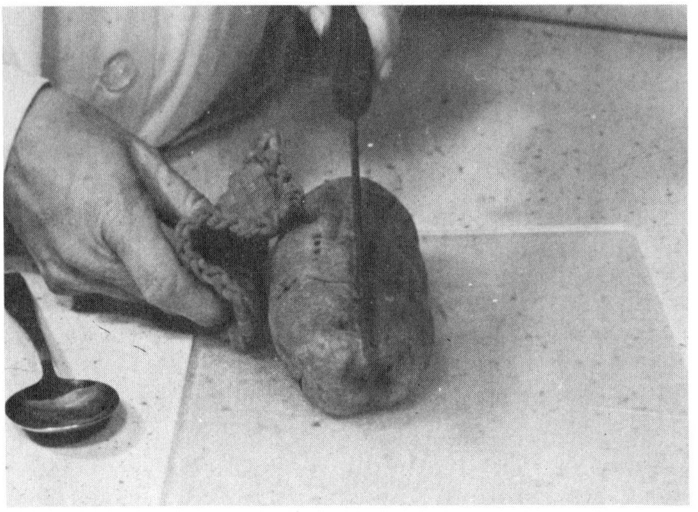

*Figure 12.7. Stuffed baked potatoes are baked until tender and then cut in half in preparation for stuffing.*

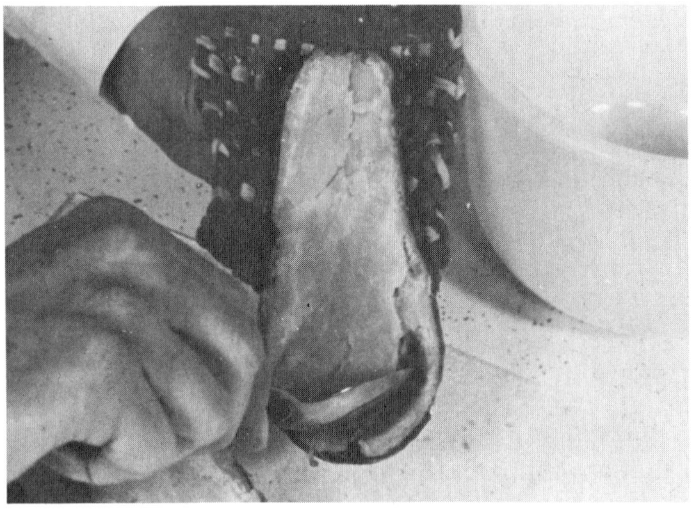

*Figure 12.8. The soft, baked potato is scooped carefully from the shells and placed in a bowl for mashing.*

*Figure 12.9. The empty shells are crisped in the oven while the interior mixture is being mashed.*

*Figure 12.10. The addition of a small amount of baking powder helps to leaven the mashed potato mixture.*

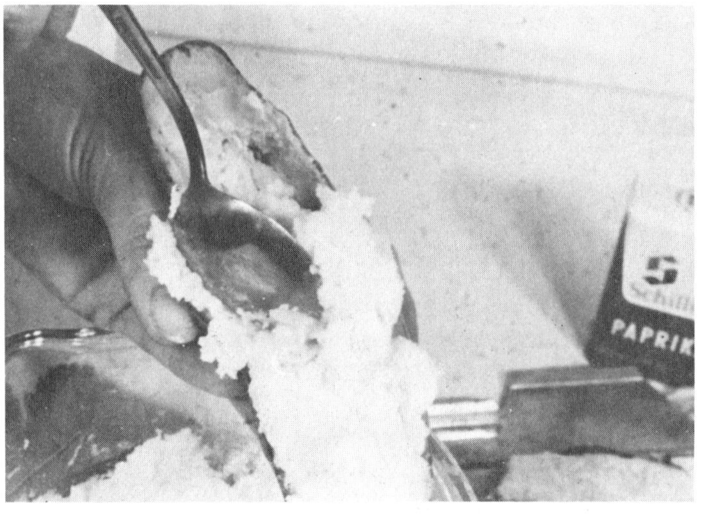

*Figure 12.11. The mashed potatoes are spooned lightly into the shells. These may be returned to the oven for reheating prior to serving, or they may be cooled and frozen.*

2. Mash potatoes, onion, sour cream, and cottage cheese together until light and fluffy. Add seasonings to taste and beat in the baking powder.
3. Allow mashed potatoes to cool before packing firmly into a plastic freezer container.
4. Leaving a half inch headspace, close the container firmly, label, and freeze.

*Note:* To serve, thaw completely and then bake at 350°F for 30 minutes or until warm throughout. This casserole also can be thawed and reheated to serving temperature in a microwave oven. If only enough will be needed at one time to serve four or fewer, pack casserole into containers of an appropriate size.

**Tater Peel Crispies**

Shells from baked potatoes (Mashed Potato en Casserole)
Melted butter or margarine

1. Cut empty potato shells into long strips (in quarters for toast or rolls, in eighths or thinner for snacks).
2. Place on cookie sheet lined with aluminum foil.
3. Brush all sides of peels with melted fat.
4. Bake at 450°F for 5 to 15 minutes until crisp and dry. Toast under broiler briefly, if desired.
5. Cool to room temperature before packaging in plastic bags, sealing, labeling, and freezing.

*Note:* Reheat at 450°F or under broiler until crisp and hot just before time to serve these as an accompaniment to a tomato juice appetizer or snack item.

**Tasty Tomato Cocktail**

1 can tomato or vegetable juice cocktail
Pimiento-stuffed olives

1. In each section of empty ice cube trays, place a drained, stuffed olive.
2. Pour the vegetable or tomato juice into each section over the olive.
3. Freeze the cubes of cocktail until completely solid before removing from the trays.
4. Immediately repackage the cubes in a plastic bag, seal tightly, label, and return to the freezer.

*Note:* Serve by placing one of the cubes in each glass. Pour seasoned tomato or vegetable juice (room temperature) over the cube. Serve as soon as the cube is melted partially, and the drink is chilled. For large glasses, use two cubes.

## MEATS, POULTRY, AND FISH

### Procedures for Freezing

Meats, poultry, and fish to be frozen need to be handled quickly and under very sanitary conditions when being prepared for freezer storage. Work areas and the operator's hands need to be very clean. Speed is also of importance so that there will be minimal opportunity for growth of microorganisms. The cuts, whole bird, or fish should be wrapped tightly with heavy aluminum foil or freezer wrap to exclude as much air as possible from the package. Seams of the package are folded with the two sides being folded down together until they firmly contact the cut. Careful sealing and accurate labeling are done prior to freezer storage. The same precautions are used in freezing cooked meats as are used for freezing raw ones. However, the cooked meats should be chilled, uncovered, in the refrigerator before packaging and freezing. This is done to avoid warming the freezer compartment unduly when the packages are added to the freezer.

### Using Frozen Meats, Poultry, and Fish

The thawing of meat and related products for use following freezer storage ordinarily is best done in the refrigerator where the temperature of the thawing food has no possibility of entering the danger zone (above 40°F), and the microorganisms have very little opportunity for growth prior to cooking. However, refrigerator thawing is very slow and requires planning as much as three days in advance of use of the food. Since one of the merits of frozen foods is to have food available to meet unexpected needs, there may be need for a more flexible approach to thawing. A microwave oven equipped with a thawing cycle provides an excellent solution to this problem. Such an oven has a special setting which cycles the magnetron tube off and on intermittently to permit quick thawing with intervening time periods to provide the opportunity for thawing to proceed more uniformly via conduction within the cut.

Thawing can be hastened even without the use of a microwave oven, when necessary. This is done by placing the cut (securely enclosed in a plastic bag) in water which is changed frequently. The water conducts warmth to the cut more efficiently than does the ambient air. The faster rate of thawing allows the meat to thaw more quickly and reduces the period available for growth of microorganisms at room temperature.

Cuts can be cooked from the frozen state if absolutely necessary. However, the large temperature differential between the cooking medium and the interior of the cut leads to overcooking the outer portions before the interior is done. For persons who like beef well done on the exterior and quite rare in the middle, this solution may be ideal. When preparing pork, this may be hazardous because of the difficulty of heating the interior to 170°F before the exterior is too dry and tough. Thin cuts and ground meats may be prepared satisfactorily from the frozen state, particularly when the thin cuts have been packaged individually or in a manner which allows easy separation.

### Special Recipes

**Freezer-Ready Beef Stew**

2 pounds beef stew meat (or leftover pot roast)
1 package mushroom gravy mix
1 package onion soup mix
8 carrots
8 potatoes (waxy, such as Red Triumph)
8 medium onions

1. Place meat and both mixes in pressure saucepan; add water to almost cover meat.
2. Following manufacturer's directions for operation, bring to 15 pounds pressure and maintain pressure

for 20 minutes. Remove from heat and allow pressure to drop gradually (about 15 minutes required).
3. While pressure is dropping, pare the carrots and potatoes, peel the onions, and cut the vegetables in bite-sized cubes.
4. Remove cover of pressure saucepan, add the vegetables, and stir in. Replace the cover.
5. Again heat to 15 pounds pressure and process for 3 minutes at 15 pounds.
6. Allow pressure to reduce gradually. Serve part of the stew when pressure has dropped and freeze the leftovers in convenient-sized containers, or freeze the entire batch, if preferred.

*Note:* Dumplings can be made by tossing 2 cups of biscuit mix with ½ cup chopped parsley, stirring in ¾ cup milk with a fork, and then dropping from a spoon onto the surface of the stew after the frozen stew has been reheated to serving temperature. The dumplings are heated for 15 minutes with the cover on the pan of stew and the heat adjusted to simmering.

**Freezer Burgers**

1 pound ground beef
1 envelope onion soup mix
½ cup finely crushed cracker crumbs

1. Mix ingredients together thoroughly, adding a tablespoon of water if mixture seems too dry.
2. Shape into 8 patties, each about 4½ inches in diameter.
3. Wrap each patty individually in a single layer of aluminum foil.
4. Label and freeze patties.

*Note:* About 20 minutes before serving, preheat oven to 425°F. Place the frozen, foil-wrapped patties on a cookie sheet. Bake in the preheated oven for 10 to 15 minutes (still wrapped in foil). Serve on toasted buns. If patties are shaped differently than described above, the baking time will be altered.

**Old Fashioned Swiss Steak**

2 pounds round steak, 1 inch thick
Flour
Seasoned salt and pepper
Oil to coat bottom of pan
1 onion, chopped fine
½ cup water
1 tablespoon horseradish

1. Cut steak into 8 portions.
2. In a bag combine enough flour, salt, and pepper to coat the meat. Shake the meat in the bag until well coated with the flour mixture. After meat has stood a few minutes, shake it in the flour mixture again.
3. Place oil in a pressure saucepan, add the meat, and brown well on both sides.
4. In a blender, combine the onion, water, and horseradish. Whirl together to make puree.
5. Spread pureed mixture over the surface of the browned meat in the saucepan.
6. Pressurize the saucepan and process for 15 minutes at 15 pounds pressure. Begin timing when the 15 pounds pressure is first reached.
7. Allow the saucepan to reduce pressure gradually when the processing time has been completed.
8. Make gravy from the drippings after the meat has been removed, if desired.
9. Serve part of the Swiss steak at this point or freeze the meat in packages of appropriate size after the meat has cooled.

*Note:* If preferred, the steak may be braised in a covered skillet on top of the range or in the oven (set at 300°F) until fork tender. Approximately 2 hours will be needed for this. Enough water will need to be added prior to braising to provide approximately ⅛ inch of water in the skillet. The level should be checked occasionally, and water added as needed to maintain this level.

**Corn Dogs**

10 weiners
Flour
1 egg
1 cup pancake mix
½ cup yellow cornmeal
1 tablespoon sugar
⅔ cup water
Oil for deep fat frying

1. Heat the weiners in boiling water just long enough to heat them through. Drain.
2. Roll each weiner in flour to coat.
3. Beat the egg in a pie plate. Blend in the pancake mix, cornmeal, sugar, and water. Allow batter to stand 15 minutes to absorb the liquid and thicken.
4. If desired, wooden skewers may be inserted in the end of each weiner.
5. Dip each weiner in the batter to completely coat the weiner.
6. Fry in deep fat which has been preheated to 375°F until brown on each side. Drain on paper towels.
7. Cool before wrapping each weiner individually in aluminum foil.
8. Label and freeze.

*Note:* Reheat for serving by placing on a cake rack positioned over a cookie sheet and baking at 400°F until defrosted, hot in the middle, and crusty.

**Freezer Chicken Barbecue**

1¼ cups all purpose flour
2 teaspoons sugar
½ teaspoon salt
3 teaspoons baking powder
1 egg
1 cup milk

2 tablespoons salad oil
2 frying chickens, cut up
Barbecue sauce

1. Beat the egg; add the milk and oil and blend well.
2. Sift the flour, sugar, salt, and baking powder together to blend thoroughly.
3. Pour the liquid ingredients into the dry ingredients all at once. Beat with an egg beater until smooth.
4. Wash chicken pieces carefully. Pat dry with paper towels.
5. Dip each piece of chicken into batter and fry in deep fat preheated to 350°F. Fry about 3 pieces of chicken at a time so the fat will recover and maintain its frying temperature.
6. Fry each piece until it is a pleasing brown and the flesh is cooked by the bone.
7. Drain on paper towels and cool before wrapping in aluminum foil or plastic bag. For convenience, separate the individual pieces with foil. Label and freeze.
8. When ready to prepare the barbecue at another time, preheat the oven to 450°F or prepare a charcoal fire.
9. Coat each piece of frozen chicken liberally with the barbecue sauce. Reheat the chicken until the sauce becomes an appealing, dark brown and the chicken is thawed and hot throughout.

**Sherried Chicken**

1 frozen, deep-fat fried, batter-dipped chicken, in serving pieces
1 can (10½ oz.) cream of chicken soup, undiluted
⅓ cup dry sherry wine
2 teaspoons dried minced onion
Lemon pepper
Sesame seeds

1. Place unwrapped pieces of frozen chicken in a shallow baking dish.
2. Combine undiluted soup, wine, onion, and a generous sprinkle of lemon pepper, mixing well.
3. Spoon sauce over chicken pieces, and sprinkle with sesame seeds.
4. Bake at 350°F for 45 minutes until sauce is bubbling and the chicken is heated to serving temperature.

*Note:* The chicken for this recipe can be prepared for freezer storage according to the directions for batter coating and deep fat frying presented in the preceding recipe. An alternative is to purchase commercially frozen chicken which has been batter-dipped and fried.

**Mushroom Chicken**

1 frozen, deep-fat fried, batter-dipped chicken, in serving pieces
1 can (10½ oz.) cream of mushroom soup, undiluted
3 ounces grated sharp cheddar cheese
½ cup slivered almonds

1. Place unwrapped pieces of frozen chicken in a shallow baking dish.
2. Spread the undiluted soup over the chicken.
3. Bake at 350°F for 30 minutes.
4. Remove briefly from oven to sprinkle grated cheese and almonds on the surface.
5. Return to oven and bake another 15 minutes until cheese has melted and the soup and chicken are well heated.

## ADDITIONAL DIMENSIONS

### Frozen Platters

Individual platters or meals to serve one can be prepared efficiently in the home to meet special meal schedules or needs of various family members. These may be prepared by arranging the foods on an aluminum tray or other freezer-worthy container. Such platters may be composed of foods remaining from a meal; these are a particularly economical idea when families eat most of the food prepared for a meal. The frozen dinner will be a welcome convenience when a family member is unable to eat with the others at some future date. In other instances, it may be practical to plan to set aside a block of time to prepare several specific dinners for future storage.

The meals provided by these platters should embody the same principles used in planning other meals. The color, shape, flavor combinations, textures, and nutritive value all need to be considered in planning the foods to combine in the platter. Usually the platter should contain a protein-rich item, at least one vegetable, and often either rice, potato, or other starch-containing product. Such menus provide excellent nutrition, particularly when rounded out by a crisp salad and a glass of milk. An example of such a menu, using recipes contained in this chapter, would include Old Fashioned Swiss Steak, Copper Pennies, and Perfect Twice-Baked Potato.

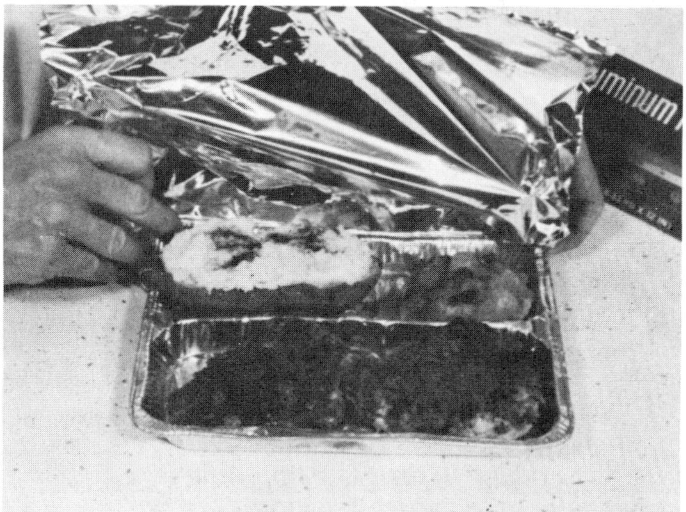

*Figure 12.12. Home prepared frozen dinners are a great convenience for hectic meal schedules and an excellent way of utilizing leftovers.*

## Frozen Desserts

**Hard-to-Believe Sherbet**

2 packages (3 oz. each) fruit-flavored gelatin
2 cups boiling water
2 trays of ice cubes

1. Dissolve gelatin in boiling water. Stir carefully and thoroughly until all gelatin is dissolved.
2. Add ice cubes and stir slowly to help cool the mixture uniformly.
3. When mixture begins to congeal slightly, remove any remaining ice cubes, and pour into a 1-quart shallow pan.
4. Cover pan tightly with aluminum foil and place in freezer after labeling.

*Note:* To serve this frozen gelatin, place the amount estimated to be needed for a meal in the small bowl of the electric mixer and beat until very smooth. Serve immediately in sherbet glasses.

**Cream Cheese Freeze**

2 packages (3 oz. each) cream cheese, softened
⅔ cup sugar
2½ teaspoons vanilla
2 cups half and half (light cream)

1. Beat the cheese, sugar, and vanilla together to soften before adding the unbeaten cream.
2. Mix the cream with the other ingredients until perfectly blended and smooth.
3. Transfer to a metal loaf pan and freeze until firm.
4. Break frozen mixture into pieces; beat with electric mixer until smooth.
5. Return to loaf pan, cover tightly, and store in freezer. Be sure the food is clearly labeled.

*Note:* This freeze may be served like any ice cream, plain, or with berries or a sauce on top.

**Frosty Pumpkin Pie**

2 baked 9-inch pie shells
1 quart vanilla ice cream
¼ cup chopped crystallized or preserved ginger
2 cups sugar
2 cups (1 lb.) canned pumpkin
2 teaspoons pumpkin pie spice
½ teaspoon salt
1 cup chopped walnuts or pecans
1 pint whipping cream

1. Place ice cream and very finely chopped ginger in mixer bowl and blend well before spreading in the baked pie shells.
2. Place in freezer until frozen solid.
3. Blend the sugar, pumpkin, spice, salt, and nuts together.
4. Whip the cream until stiff.
5. Fold the pumpkin mixture into the whipped cream until mixture is homogeneous.
6. Spread pumpkin mixture in a smooth layer over the ice cream.
7. Cover with plastic wrap before placing in freezer for storage.

*Note:* For ease in serving, remove these pies from the freezer about 5 minutes before the pie will be served.

**Pumpkin Ice Cream Pie**

1 baked 9-inch pie crust
1 quart vanilla ice cream
2 cups (1 lb.) canned pumpkin
¼ cup sugar
1 teaspoon ground cloves
1 teaspoon ground cinnamon
1 teaspoon ground nutmeg

1. Place the vanilla ice cream, pumpkin, sugar, and spices in a large mixer bowl, and beat on an electric mixer until completely blended.
2. Pile mixture high in the baked pie shell.
3. Place in freezer until firm before removing it from the freezer just long enough to cover tightly with plastic wrap.

*Note:* These versions of pumpkin pie are particularly convenient to serve for holiday meals because they can be prepared well in advance and then be removed from the freezer about 5 minutes before being served.

**Never-Fail Meringue**

1 9-inch baked pie shell
1 quart ice cream (any flavor)
3 egg whites
½ teaspoon cream of tartar
1 cup marshmallow whip topping
1 teaspoon vanilla

1. Soften the ice cream just enough so that it can be spread to fill the pie shell. Spread ice cream with a spatula to make a uniform thickness in the shell.
2. Beat egg whites to the foamy stage with an electric mixer at the highest setting.
3. Add cream of tartar and continue beating until soft peaks form.
4. Add marshmallow whip, a spoonful at a time, beating well after each addition.
5. Continue until all of the marshmallow has been added. Then beat in vanilla.
6. Pile meringue on the ice cream pie and spread with a spatula to seal meringue to edges of the pie.
7. Freeze meringue-topped pie.

*Note:* To serve this pie, preheat the oven to 450°F before placing the pie in the oven. Bake for 3 to 5 minutes until the meringue peaks are browned delicately. Serve immediately. This is a pastry version of a Baked Alaska with the unique convenience of freezer storage.

**Do-Ahead Roast Turkey**

1 turkey
1 pkg. (13 oz.) dressing mix
1 cube margarine
2 medium stalks celery and leaves, chopped
1 small onion, chopped

1. Roast unstuffed turkey according to personal preference or using standard cookbook recipe.
2. Simmer giblets and neck (if desired) in a covered saucepan with water to cover. Add a stalk of celery and a coarsely chopped onion as optional ingredients.
3. When giblets are very tender, remove from the stock and cut in small pieces. Save the stock from cooking the giblets to use in making the dressing and gravy.
3. Saute the celery and onion in the margarine until slightly translucent.
4. Mix the vegetables and margarine with the dressing.
5. Add enough of the stock to moisten the dressing to the point where the dressing can be formed into a ball which will fall apart when touched.
6. Place dressing in a metal pan (8 x 8"), cover tightly with foil, and bake for an hour at 350°F.
7. Remove the foil, and place the pan of dressing in the refrigerator to chill thoroughly. Place the chilled dressing in a freezer container, label and freeze for later use with the turkey.
8. As soon as the turkey is roasted, cool as efficiently as possible, preferably in the refrigerator. When cooled completely, place the whole turkey in a large plastic oven-browning bag, seal tightly, label and freeze.
9. While turkey is cooling, make gravy from the drippings and stock. Cool and then place in plastic container for freezer storage.

*

10. Thaw the roasted turkey in the refrigerator, allowing a day or two, depending on the size of the turkey. The day of the dinner, thaw the gravy and dressing in the refrigerator.
11. No later than two hours before serving time, transfer the turkey in its bag to a large roasting pan, and make 4 holes in the top of the bag before beginning to reheat the turkey in a 300 to 350°F. oven. At the end of two hours, turn the oven to 150°F. to hold until ready to serve. The dressing is reheated, tightly covered with foil, for the final hour that the turkey is being reheated. The gravy is reheated in a pan on the range at the time final arrangements are being completed for dinner.

**Frozen Whipped Cream Dollops**

Whipping cream
Powdered or granulated sugar
Vanilla

1. Beat whipping cream until stiff.
2. Sweeten and flavor as desired.
3. Spoon attractive dollops of whipped cream onto aluminum foil which has been placed on a cookie sheet.
4. Place the sheet of dollops in the freezer and freeze solid.
5. Remove from the freezer long enough to package frozen dollops in a freezer carton or plastic bag.
6. Label and store in freezer after sealing.

*Note:* This technique of storing frozen dollops of whipped cream can be used any time small amounts of whipped cream are left after serving desserts. The dollops are served simply by placing them on the dessert to be garnished just shortly before serving.

# Glossary

**Achromobacter** – Cryophilic bacteria sometimes found in dried and frozen fruits and vegetables.

**Aerobacter** – Psychrophilic bacteria sometimes occurring in frozen and dried fruits and vegetables.

**Aerobes** – Organism, such as a bacterium, which requires oxygen for survival.

**Albedo** – The white portion of the peel of citrus fruits; portion is rich in pectin and is utilized commercially as a source of pectin.

**Alcaligenes** – Cryophilic bacteria sometimes found in frozen fruits and vegetables.

**Alliinase** – Enzyme in onions and other members of the allium family; converts cysteine sulfoxide and its derivatives into allium compounds which yield the characteristic odor and flavor of these vegetables.

**Allium** – Any of the genus of bulbous herbs of the lily family; includes garlic, onions, chives, leeks, and shallots.

**Amylopectin** – The branched-chain fraction of starch; glucose units linked by 1,4- and 1,6-alpha glucosidic linkages to form the polysaccharide.

**Amylose** – The straight-chain fraction of starch; glucose units linked by 1,4-alpha glucosidic linkages to form the polysaccharide. This fraction is needed for gel formation to occur.

**Anaerobe** – Organism, such as a bacterium, which does not require oxygen for survival.

**Anthocyanin** – Flavonoid pigments which contribute reds and purples to the color of fruits and vegetables.

**Anthoxanthin** – Flavonoid pigments which contribute white to colorless appearance to fruits and vegetables.

**Antioxidant** – Compound which oxidizes very readily, thus preventing oxidation of other compounds which interact less freely with oxygen.

**Appert, Nicolas** – Frenchman who sometimes is referred to by the title of "Father of Canning".

**Ascorbic acid** – Also called vitamin C. This vitamin is used to prevent browning of fruits because it takes up oxygen very readily, thus protecting the fruit color.

**Bacillus anthracis** – Bacteria, the spores of which are destroyed readily by heat.

**Bacillus coagulans** – Facultative anaerobe which is capable of producing acid unless destroyed by heat processing; causes flat-sour spoilage, particularly in tomatoes.

**Bacillus macerans** – A facultative anaerobe which sometimes causes flat-sour spoilage in canned fruits.

**Bacillus polymyxa** – Facultative anaerobe which can cause flat-sour spoilage in canned fruits.

**Bacillus stearothermophilus** – Facultative anaerobe capable of causing flat-sour spoilage even in low-acid foods because it forms thermophilic spores.

**Bacillus subtilis** – Spore-forming bacteria which are destroyed in 20 minutes at 100°C.; capable of causing food-borne illness in thawed, frozen foods which are not refrigerated properly.

**Blanching** – To scald or parboil in water or steam. Process commonly used in preparing vegetables for freezing or drying; inactivates enzymes.

**Botulism** – Food-borne illness (often fatal) caused by ingestion of the toxin produced by *Clostridium botulinum.*

**Brine** – Salt solution; used in preparing brined pickles.

**Browning** – Darkening of the flesh of fruits which may occur as the result of reaction between amino acids and sugar or enzyme action.

**Butylated hydroxyanisole** – Antioxidant commonly used in foods produced commercially. Also called BHA.

**Butylated hydroxytoluene** – Antioxidant frequently used in commercial foods. Also called BHT.

**Byssochlamys fulva** – A mold capable of causing disintegration of the pectic substances in fruits; action produces mushy texture.

**Carotenoids** – Orange pigments contained in yellow and orange vegetables.

**Catalase** – Enzyme occurring in cauliflower and other vegetables; causes detrimental changes during frozen storage unless inactivated by blanching.

**Cellulose** – Polysaccharide which is a principal constituent of cell walls. Composed of glucose units linked linearly by the 1,4-beta glucosidic linkage.

**C-enamel** – Also called corn-enamel. Lining for metal cans, designed specifically to meet the packaging requirements for canned corn and hominy.

**Checking** – Breaking the skin of fruits in preparation for drying. A lye solution can be used to effect these breaks, but a fork can be used equally well to accomplish the perforations needed for facilitating drying.

**Chlorophyll** – Green pigment in fruits and vegetables. Undergoes change to pheophytin when the magnesium ion is replaced by hydrogen, as happens during the prolonged heating period required in canning.

**Chloroplast** – Structure in the cytoplasm of parenchyma cell where the chlorophyll is found.

**Chromoplast** – Structure in the cytoplasm of the protoplast of the parenchyma cell where carotenoid pigments are found.

**Chuno** – Dried potatoes prepared by natives living in the Andes many centuries ago. Prepared by walking on the potatoes to make a pulp before drying.

**Cladosporium** – Cryophilic mold which grows at freezer temperatures.

**Clostridium bifermentans** – A mesophilic obligate anaerobe.

**Clostridium botulinum** – Spore-forming, mesophilic anaerobe; capable of producing highly poisonous toxin. Types A and B are very heat resistant.

**Clostridium histolyticum** – A mesophilic obligate anaerobe.

**Clostridium nigrificans** – A thermophilic obligate anaerobe which produces hydrogen sulfide if not destroyed during canning. Iron in a can will react with the hydrogen sulfide to cause black food products.

**Clostridium pasteurianum** – Butyric anaerobe which is particularly troublesome in canned foods at a pH between 4.0 and 4.5.

**Clostridium putrificum** – A mesophilic obligate anaerobe.

**Clostridium sporogenes** – A putrefactive anaerobe which may cause spoilage in canned foods, particularly those with a pH of 4.5 or higher.

**Clostridium thermosaccharolyticum** – A thermophilic anaerobe capable of producing large quantities of carbon dioxide and hydrogen; causes spoilage referred to as "swell" and a cheese-like odor.

**Cold pack** – Packing of raw food into jars in preparation for canning. The liquid poured over the raw food may be boiling water, sugar syrup, or broth.

**Collenchyma cells** – Elongated, thick-walled cells contributing to the fibrous quality of vegetables, such as celery.

**Conduction** – Transmission of energy via a conductor. A method by which heat is transferred throughout food masses being canned.

**Conserves** – Preserves containing chopped nuts.

**Continuous phase** – The liquid surrounding the droplets in an emulsion.

**Convection** – Transfer of heat energy by circulation of a liquid, as occurs in the canning of foods packed in a liquid.

**Corn enamel** – Lining for metal cans, designed specifically to meet the packaging requirements for canned corn. Also called C-enamel.

**Cysticercus bovis** – Larval stage of beef tapeworm.

**Dehydrator** – Apparatus designed for drying foods. Usually is equipped with a fan and a thermostatically controlled heat source.

**Demethylation** – The removal of methyl ($-CH_3$) groups from a chemical compound. Occurs when pectin is transformed into pectic acid as fruit progresses from ripe to overripe.

**Denaturation** – Change in protein from the native state to an altered state (usually less soluble). Often is caused by applying heat to protein.

**Dial gauge** – Gauge used in some pressure canners to indicate the actual amount of pressure inside the canner.

**Endamoeba histolytica** – Parasite causing amoebic dystentery. Freezing for at least 24 hours at a temperature of -18°C. (0°F.) kills this parasite if it is present in foods.

**Escherichia coli** – Bacteria often occurring in food. This is the most common of the *Escherichia*, and protected by the presence of sugar, but detrimentally influenced by salt.

**Exhausting** – Process of evacuating the air as much as possible from a pressure canner by heating with the valve open for several minutes after steam is coming continuously from the petcock.

**Extraction** – Process of removing the juice and the pectin from fruits in preparation for forming pectin gels.

**Facultative anaerobe** – Thermophilic, spore-forming bacilli capable of causing flat-sour spoilage in canned foods if not killed during processing. Examples of these bacteria which do not require oxygen include *Bacillus stearothermophilus*, *Bacillus coagulans*, *Bacillus macerans*, and *Bacillus polymyxa*.

**Flat-sour spoilage** – Spoilage in canned foods in which acid is produced by microorganisms (*B. stearothermophilus*, for example) without accompanying production of gas.

**Flavobacterium** – Example of a bacteria classified as cryophilic. May cause spoilage in frozen foods.

**Freezer burn** – Desiccation or dehydration of the surface of a frozen food; caused by a faulty covering of frozen foods during storage. Superficial surface damage occurs and is particularly apparent in poultry or meats when a tear in their packaging occurs.

**Fructose** – Monosaccharide; one of the two breakdown products of sucrose.

**Galactose** – Monosaccharide; one of the two breakdown products of lactose (milk sugar).

**Galacturonic acid** – Organic acid derivative of galactose; the acid radical is on the sixth carbon, i.e., the carbon outside the ring structure of galactose. When esterified with methanol, long polysaccharide chains of this compound are called pectin or pectinic acid. They are utilized in making jams and jellies.

**Gelatin gel** – Colloidal dispersion which holds its shape because of the cross linking of the protein molecules (gelatin) it contains.

**Gelatinization** – Imbibition of water into starch granules in the presence of heat and water. The physical swelling of the granules, the leaching of amylose from the granule, and the hydrogen binding of water in the granule are changes leading to increased viscosity and clarity of the hot starch paste.

**Gel** – Colloidal dispersion in which a solid is dispersed in a liquid, with the solid providing a structural network which traps the liquid and causes the dispersion to hold its shape without flowing.

**Glucose** – Monosaccharide; one of the breakdown products found when any of the common disaccharides are hydrolyzed. Also provides the structural units for starch and cellulose.

**Glucoside** – Compound which yields glucose upon hydrolysis.

**Glycogen** – Animal counterpart of the starch found in plants. The storage form of carbohydrate in animals. Composed of glucose units linked with 1,4-alpha glucosidic linkages.

**Golgi bodies** – Discrete part of the cytoplasm of parenchyma cell.

**Headspace** – Space left at the top of a jar or freezer package to allow for expansion during heat processing or freezing of foods.

**Hemicelluloses** – Polysaccharides in plant cell walls which are less complex than cellulose. Can be hydrolyzed to monosaccharide.

**Hollow pickles** – Pickles failing to have the desired crisp interior; caused by using cucumbers of poor quality, either poorly developed or stored too long.

**Hominy** – Hulled corn with the germ removed; usually processed in lye.

**Hot pack** – The filling of jars for canning using hot, cooked food. Method often is used for canning meat, fish, poultry, pickles, and relishes. Shrinkage is minimized by this process.

**Hydrogen bonds** – Secondary bonding forces between oxygen and hydrogen. Of importance in gelatinization and gel formation as well as in many other facets of food preparation.

**Hydrogen sulfide** – Gas with a distinctive, strong odor. Can be produced by the action of viable *Cl. nigrificans* in canned foods. Sometimes will cause bulging of can lids.

**Hydrogen swells** – Spoilage of canned food in which considerable hydrogen gas is produced by viable thermophilic anaerobes.

**Hydrolysis** – Chemical reaction in which a complex compound is broken into smaller units with the uptake of a molecule of water. Often favored by the presence of acid or enzymes.

**Hydrolytic rancidity** – Development of rancidity in fats due to the freeing of fatty acids from the fat molecule, a reaction facilitated by lipase (enzyme catalyzing lipolytic reactions).

**Hydroxyl group** – Functional group commonly occurring in food compounds. Consists of an oxygen and a hydrogen atom (-OH).

**Iodized salt** – Sodium chloride to which potassium iodide has been added as an important nutritional additive. However, iodized salt used in making pickles will cause the pickles to darken.

**Isothiocyanates** – Desirable flavoring components developed in cabbage and related compounds. Formation is catalyzed by enzymes when the vacuole and protoplast come in contact.

**Jams** – Pectin gels containing pieces of fruit.

**Jellies** – Pectin gels made from fruit juices, but with no pieces of the fruit pulp being present.

**Jelmeter** – Device used to measure the viscosity of pectin-containing fruit juice to determine the amount of sugar required to produce a pectin gel of optimum quality.

**Jerky** – Long, thin pieces of meat dried either in the sun or artificially in the oven or dehydrator. Often meat to be jerked is soaked in a marinade first to enhance the flavor.

**Lactic acid** – Organic acid produced by lactic-acid bacteria in milk.

**Lactose** – Disaccharide in milk. Also called milk sugar; composed of a unit of glucose and one of galactose linked together with the elimination of a molecule of water.

**Leathers** – Dried fruit or vegetable slurry.

**Leucoplast** – Part of the cytoplasm of a parenchyma cell where the starch granules are deposited.

**Lipase** – Enzyme catalyzing hydrolysis of fats.

**Lipoxidase** – Enzyme catalyzing oxidation of fats.

**Low-acid food** – Food with a pH above 4.5. Requires pressure canning to achieve sufficiently high temperatures to kill spores of microorganisms which can flourish at this high pH.

**Lye solution** – Solution used to check the skins of fruits to facilitate drying. Made by dissolving a tablespoon of lye in two quarts of cold water.

**Lysosomes** – Organelles in the cytoplasm of the protoplast in which hydrolytic enzymes are found.

**Maillard reaction** – Reaction between amino acid and sugar which results in browning.

**Mason jar** – Jar designed by J.L. Mason for the canning of foods. Its special feature allows some evacuation of the jar during processing and then air-tight sealing.

**Melanin** – Dark brown pigment in plants.

**Mesophilic obligate anaerobes** – Obligate microorganisms which can thrive at moderate temperatures in the absence of oxygen. *Cl. botulinum* is perhaps the best-known example. Several other microorganisms (*Cl. putrificum, Cl. histolyticum, Cl. bifermentans, Cl. sporogenes,* and *Cl. pasteurianum)* are examples.

**Micrococcus** – Example of a cryophilic bacterium which may cause food spoilage in frozen foods.

**Microsome** – Part of the cytoplasm of the protoplast where protein may be formed.

**Middle lamella** – Cementing material of plants; contains the pectic substances.

**Mitochondria** – Bodies in cells containing fats, protein, and enzymes in abundance; centers for cellular respiration.

**Monilia** – Cryophilic mold which may grow on food during frozen storage.

**Obligate aerobes** – Microorganisms which require oxygen for their growth. Those found in food are destroyed by heat, hence do not present a hazard in canned foods.

**Obligate anaerobes** – Microorganisms which can grow in the absence of oxygen. May be either mesophilic or thermophilic.

**Open kettle canning** – Method of canning in which the food is heat treated prior to being placed in sterile containers. Satisfactory for making jams and jellies because their high sugar content is a suitable protection. Not recommended for any other type of canning.

**Osmotic pressure** – Pressure created by having unequal concentrations of a solution on either side of a semipermeable membrane; water will flow from the side where the solute concentration is low to the side where it is high, in an attempt to equalize the solution on both sides of the membrane. Causes swelling or dehydration of cells, depending on the direction of flow of the water.

**Oven canning** – Old-fashioned method sometimes used in the past for heat processing of foods. Method is definitely not recommended because it is not possible to heat the foods to a high enough temperature to assure killing of harmful microorganisms which may be present. Foodborne illnesses and food spoilage are possible consequences of oven canning.

**Oven drying** – Drying of food by placing thin slices of food in the oven and allowing it to become dehydrated as a result of operating the oven at a low temperature (usually 140°F.).

**Oxidative rancidity** – Detrimental uptake of oxygen by fats. Ordinarily occurs as the result of the contact of oxygen with double bonds (unsaturated fatty acids) and catalyzed by lipoxidases.

**Parenchyma cell** – Cells commonly found in the edible parts of plants. Characteristically have 14 sides and strong cell walls. Composed of the protoplast, vacuole, and cell wall.

**Pectic acid** – Form of pectic substance found in overly ripe fruit; has lost its ability to form a gel.

**Pectic substances** – Polysaccharides found in the middle lamella of plant cells. Polymers of galacturonic acid.

**Pectin** – Pectic substance capable of forming a gel. Methylated polymer of galacturonic acid.

**Pectinic acid** – Synonym for pectin.

**Pemmican** – Jerky made from meat (usually deer or buffalo), fat, and sometimes molasses and flour.

**Penicillium** – Cryophilic mold which is capable of growing during frozen storage.

**Peroxidase** – Enzyme capable of catalyzing formation of peroxides in fats, leading to the development of oxidative rancidity in fats.

**Petcock** – Valve which can be opened and closed to control air flow in a pressure saucepan or canner.

**pH** – Hydrogen ion concentration, expressed on a logarithmic scale. In food preservation, a pH of less than 4.5 allows foods to be processed at atmospheric pressure; lower acidity (higher pH values) requires pressure canning to achieve the higher temperatures needed to kill microorganisms which flourish in a less acidic environment.

**Pheophytin** – Yellow green pigment resulting when the magnesium ion in chlorophyll is replaced by hydrogen, as occurs in an acidic environment or with prolonged heating.

**Pickling** – Preservation of food by the addition of acid and often salt; heat processing at atmospheric pressure is used.

**Plasmalemma** – Membrane surrounding the protoplasm adjacent to the cell wall.

**Plasmodesmata** – Part of the cell wall; strands of cytoplasm interconnecting cells.

**Polygalacturonase** – Enzyme catalyzing hydrolysis of pectic substances.

**Pressure canner** – Large, heavy metal container which can be pressurized to permit heat processing at temperatures higher than can be achieved at atmospheric pressure. Essential piece of equipment for canning vegetables and meats in quantity.

**Pressure saucepan** – Special type of saucepan designed to be pressurized so that temperatures higher than can be achieved at atmospheric pressure can be reached. Device usually has a capacity no larger than 6 quarts, and is only deep enough to permit processing of pint jars.

**Primary wall** – Structural wall of parenchyma cell. Composed of pectic substances, non-cellulose polysaccharides, hemicelluloses, and cellulose.

**Protopectin** – Form of pectic substances found in immature fruit. Does not have gel-forming properties.

**Protoplast** – Portion of parenchyma cell comprising the nucleus, cytoplasm, and membranes surrounding the cytoplasm.

**Pseudomonas fluorescens** – Cryophilic bacteria protected by sugar and harmed by salt.

**Psychrophiles** – Microorganisms capable of growing at temperatures ranging from room temperature to temperatures as low as -12°C. (10°F.). Examples include *Pseudomonas*, *Achromobacter*, *Flavobacterium*, *Escherichia*, *Aerobacter*, and *Alcaligenes*. *Penicillium* is a psychrophilic mold.

**Quinones** – Phenolic compound formed during the browning of light-colored fruits.

**Rancidity** – Development of off-flavors in fats and fat-containing foods as the result of hydrolysis or oxidation; often enzyme-catalyzed.

**Raw pack** – Packing of foods, particularly vegetables, into jars for canning without heating the food at all prior to processing of the jars.

**Reprocessing** – Repeating of water bath or pressure canning procedures when jars have failed to seal properly the first time they are processed.

**Retrogradation** – Crystallization or precipitation of amylose in gelatinized starch mixtures during storage. This is a particular problem when such mixtures are frozen and thawed. Heating reverses retrogradation because it dissolves the amylose crystals.

**Ribosome** – Granule in protoplasm containing ribonucleic acid; prominent in protein synthesis.

**Salmonella** – Bacteria capable of surviving frozen storage and then resuming growth in thawed foods such as chicken, dairy products, and egg-containing dishes.

**Shigella** – Microorganism capable of causing food poisoning if food is not handled with proper refrigeration and sanitation measures after frozen storage and thawing.

**Shriveled pickles** – Poor quality of pickles due to using too much salt or sugar or holding the pickles too long before processing them.

**Slippery pickles** – Caused by growth of molds and yeasts in the pickling medium; results from the presence of cucumber blossoms, too weak a pickling solution, failure to remove scum from the brine solution, or using alkaline water.

**Sodium bisulfite** – Sulfur compound used to prevent browning in fruits being dried.

**Spore** – Resistant reproductive body produced by some bacteria and yeasts.

**Sporotrichium** – Cryophilic mold capable of growing at freezer temperatures.

**Staphylococci** – Bacteria which reproduce well in protein-containing foods, causing severe gastro-intestinal discomfort when consumed in foods. Careful handling of thawed frozen foods needs to be done to prevent growth of this common bacterial contaminant.

**Starch** – Polysaccharide composed of many units of glucose polymerized into either amylose or amylopectin, the two fractions of starch.

**Sucrose** – Table sugar; disaccharide composed of a unit of glucose and one of fructose.

**Sulfide spoilage** – Spoilage in canned foods caused by failure to kill *Cl. nigrificans*. Results in production of hydrogen sulfide.

**Swell** – Spoilage in canned foods caused by production of hydrogen gas by thermophilic anaerobes.

**Syneresis** – Drainage of liquid from a gel.

**Taffies** – Fruit or vegetable leathers.

**Thermal shock** – Abrupt change in temperature; of particular concern in handling jars being removed from a pressure canner.

**Thermophiles** – Microorganisms resistant to destruction by heating; those which reproduce well at relatively high temperatures. May be classified as facultative or obligate anaerobes, both of which may cause spoilage in improperly processed canned foods.

**Tonoplast** – Membrane of the protoplast which separates the protoplasm from the vacuole.

**Toxin** – Colloidal substance, protein in nature, which is poisonous. An example is the substance produced by *Cl. botulinum*.

**Trichinella spiralis** – Parasite sometimes found in pork. Destroyed by frozen storage.

**Trichinosis** – Condition caused by ingestion of viable *Trichinella spiralis*.

**Vacuole** – Portion of the parenchyma cell containing water, salts, amino acids, vitamins, oil droplets, and the flavonoid pigments.

**Water bath canner** – Large kettle equipped with a rack and a lid; used in processing canned foods at atmospheric pressure.

**Water bath canning** – Method of preserving fruits in a closed jar, using heat to kill microorganisms. Processing is at atmospheric pressure. Suitable only for canning acidic foods, such as fruits and pickles.

**Water blanching** – Brief boiling of foods, usually vegetables, in boiling water to inactivate enzymes. Used in preparing vegetables for drying and freezing.

**Weighted gauge** – Specially-designed weight which is used to control the pressure within a pressure saucepan or canner.

# Chemical Structures

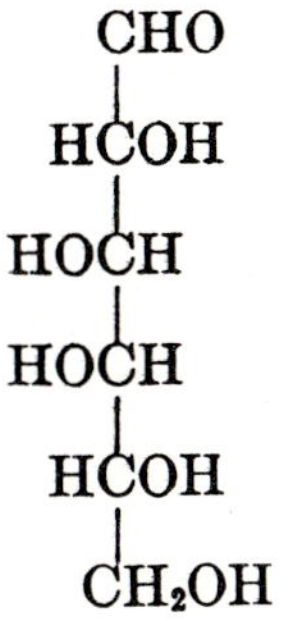


Galactose

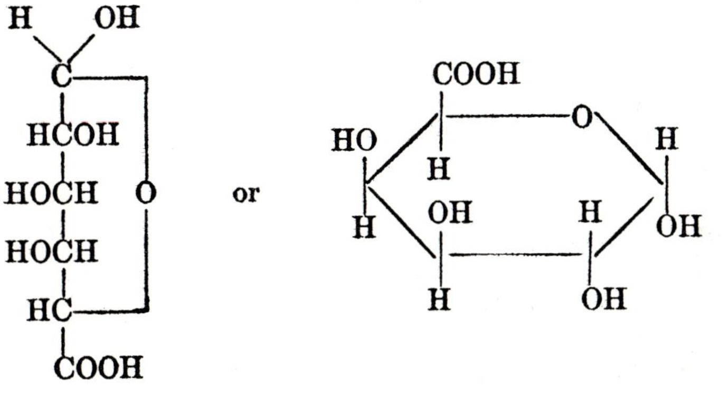


Galacturonic acid

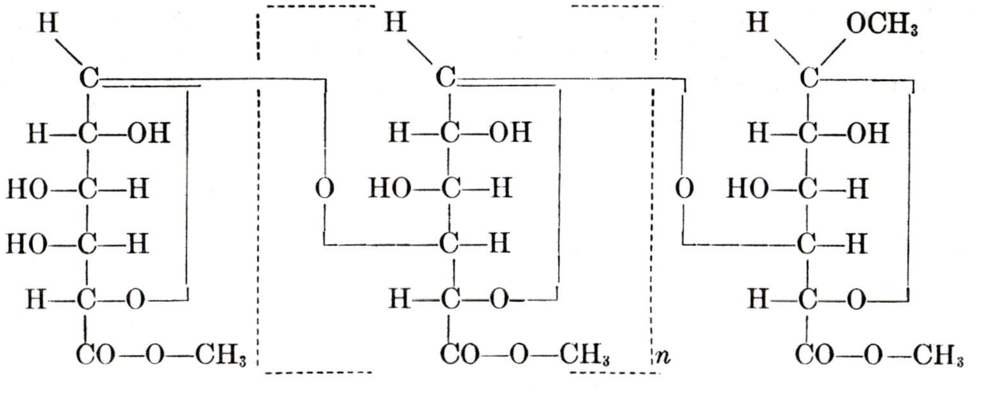


Pectin

# *Index*